MANKIND'S CHAOTIC AND MYSTERIOUS PAST

Post-Paleolithic Times (20,000 Years Ago) to the Present

Nemesis's Influence and Prophecy

DOUGLAS B. ETTINGER

Mankind's Chaotic and Mysterious Past
Copyright 2017 by Douglas B. Ettinger

Published by:
Douglas B. Ettinger

Check out the author's recently published book

The Great Deluge: Fact or Fiction?

Making Sense of and Bringing Together All the Reasonable Scientific
Hypotheses and Legends of Worldwide Cultures

I.

CONTENTS

LIST OF TABLES AND FIGURES

DEDICATION

I dedicate this book to all those wonderful women who have taken excellent care of me through the years allowing me to reach my mid-seventies. They are my grandmother, mother, aunts, sister, and nieces. They are my ex-wife, my significant other and soul mate, all my ex-girlfriends, all my secretaries in the workplace, my website director, and assistant self-publisher. My existence has been well loved and adorned; without them my life would be brutal and meaningless. I thank the Creator for his wisdom and invention of the other gender.

ACKNOWLEDGEMENTS

Unlocking the paradigm of a quiet, stable solar system with no frequent catastrophic events and no orbiting sister star is made extremely difficult. Mankind has suffered one or more 'reset buttons' that caused the sudden collapse of infrastructures which no mainstream academia accepts. My appreciation is extended to many individuals that have forged ahead in spite of being publicly ridiculed and shunned. Suppression of their research and concepts is due to the simple reason that it contradicts what has been previously taught and solidified with textbooks, government funding, and doctoral degrees. These soldiers battling against present dogma are listed here within certain categories:

CELESTIAL INTRUDERS AND DESTROYERS

Nemesis Goddess was the ancient Greek goddess of divine retribution. She meted out punishment for evil deeds, undeserved good fortune, and hubris or arrogance before the gods. Her Roman counterpart was Invidia, the goddess of jealousy as well as vengeance. Another meaning for this goddess was "the inescapable". All these personality traits are very apropos for a destructive star or planet that keeps returning to exact vengeance and punishment on mankind. Did the writers of antiquity know more about natural history than modern man realizes?

Zecharia Sitchin, writer of the *Twelfth Planet*, originated the Sun's orbiting star or planet with a period of 3600 years causing various degrees of calamity with Earth and other planets with each return. Importantly, he provided the reason for our genesis and the Great Deluge with his translations of ancient languages.

Isaac Asimov, prolific writer of non-fiction, popularized the idea that the Moon is another planet sharing Earth's orbit. This idea started the debunking of the still popular nebula hypothesis of solar system formation. Coincidentally, Asimov also wrote a fiction book, *Nemesis*, about a red dwarf star passing through our region of space.

D. S. Allan & J. B. Delair, writers of *Cataclysm! Compelling Evidence of a Cosmic Catastrophe in 9500 BC* made a very complete compilation of the world's traditions about the Great Deluge and other related cataclysms. The trigger for their events was another cosmic intruder called

'Phaeton' that passed through the solar system and collided with the Sun. Importantly, Phaeton was not the orbiting Nemesis discussed in this book.

Immanuel Velikovsky, author of *Worlds in Collision,* is one the first originators of creating the marriage of myth and science. He attempted to connect the unexpected movements of the planets with his interpretations of the legends of antiquity.

Ev Cochrane wrote an important article, "Anomalies in Ancient Descriptions of the Sun-God" that contributes immensely to my Nemesis hypothesis. Cochrane, a comparative mythologist, spectacularly listed descriptions of how the Sun-Gods of ancient Mesopotamia, Egypt, and India differ with our Sun today. These differences agreed perfectly with the envisioned Nemesis star and those periods of time when dual Suns existed in the sky.

DESTROYED AND LOST ANCIENT CIVILIZATIONS HAVING ADVANCED TECHNOLOGIES

Robert Bauval & Graham Hancock, popular fringe scientists, championed the re-dating of the Giza complex in Egypt by using their 'Orion Correlation Theory'. The theory explains the unique arrangement of the Giza pyramids which leads amazingly to a proof about the amount of Earth's crustal/mantle shell displacement during the Flood.

Robert M. Schock, a renowned geologist, published articles that re-date the Great Sphinx of Giza via water erosion; of the Bolivian Puma Punku complex via lichen and stalactite studies; and of the Easter Island half-buried Moais via soil erosion. His re-dating corroborates the Great Deluge timeline with the fall of several ancient cultures.

Eric von Daniken, author of *Chariots of the Gods*, was definitely labeled as a pseudo-scientist, but justly posed unanswered questions. He popularized ideas about past advanced civilizations and possibly extra-terrestrial astronauts creating the mysterious megalithic structures found worldwide.

DISPLACEMENT OF EARTH'S CRUSTAL/MANTLE SHELL

W. R. Farrand wrote *Frozen Mammoth and Modern Geology* and was one of the first paleontologists/geologists in 1961 to expose the enigma about mammoth bones piled in small mountains on islands along the Siberian coastline and carcasses preserved by being quickly frozen.

Charles H. Hapgood wrote *The Earth's Shifting Crust* and *The Path of the Pole* describing how the enigma of the extinct mammoths could be answered by natural causes.

John White placed the idea of shifting poles or slipping of the crust on the public radar by writing *Pole Shift* in 1980. Of course, he was encouraged by Hapgood's ideas and the recently learned periods of deglaciation, the rapid freezing of fauna, and global magnetic field reversals. Reluctantly, he provided no cause for such events.

Walt Brown, Ph.D. hypothesized an accelerated shifting of the tectonic plates and reasons for sudden burial and freezing of northern megafauna in his book, *In the Beginning – Compelling Evidence for the Flood.* He uses his 'hydroplate' hypothesis which is partially adopted by my book. Our major difference is that he uses for the opening of oceanic ridges the accumulated heat energy from tidal forces, whereas I use the trigger of a close encounter of a celestial intruder that disturbs the crust.

T. Gold, a catastrophist, has studied and published in 1962 his theory of how mega-coronal mass ejections (CME's) can maintain their magnetic field strength, reach the Earth, and cause calamity by shrinking the Earth's protective magnetosphere. These past intense solar outbursts are based on well-documented astronomical and geophysical evidence.

Robert Johnston, an independent researcher, published in a 'New Concepts in Global Tectonics' (NCGT) Journal an article, *Massive Solar Eruptions and Their Contributions to the Cause of Tectonic Uplift.* He was searching for the unknown energy required to cause the synchronous uplift of mountain ranges such as the Andes in South America. He used the model of mega-coronal mass ejections for the source. However, I accidently discovered that this energy source is not only powerful enough to displace the Earth's crustal/mantle shell, but has the magnetic field strength and an aligning vector force to rotate the shell by an already predicted amount.

CONNECTING ARCHETYPE SYMBOLS OBSERVED IN AN ANCIENT SKY WITH LABORATORY HIGH DENSITY PLASMA DISPLAYS

Anthony L. Peratt, Ph.D., Life Fellow of IEEE and member of Los Alamos National Laboratory, is a plasma physicist who wrote *Physics of the Plasma Universe.* He was one of the original discoverers that worldwide petroglyphs recorded in antiquity are associated with morphological types of plasma displays created in the laboratory. His research compares high-current plasma Z-pinches and both plasma stability and instability data with prolific petroglyphs and archetypes such as aurora funnels, stickman, big-horn-sheep profiles, and squatter man. Please refer to his website: www.plasmauniverse.info.

David Talbott & Wallace Thornhill co-authored *Thunderbolts of the Gods* and *The Electric Universe* that explained ancient archetypes or symbols as a manifestation of the ever-changing solar system viewed by earlier people of antiquity. The archetypes are then identified as electrical phenomena that are reproduced in plasma research laboratories.

Wallace Thornhill also wrote very important articles accompanied by videos titled *The Electric Comet* and *The Lightning Scarred Planet Mars* for the Electric Universe (EU) group of independent scientists. Thornhill explains how comets are only a conglomeration of rock that becomes electrically charged by orbiting into higher density solar winds and sputtering away observable plasma. He also illustrates how most of the defacement of Mars is caused by high energy electrical discharges and not impacts or erosion commonly known to occur on Earth.

Andrew Hall, an independent physicist, published blogs titled "Arc Blasts" for the Thunderbolt Project group (part of the EU group) explaining the electrical scarring of Earth. A dielectric breakdown of the atmosphere via glowing plasma figures call 'stick man' that connected the charged ionosphere with Earth's crust sputtered away rock. Anomalous geologic formations such as buttes, mesas, arches, hoodoos, and flatiron rocks were all caused by electrical phenomena per Hall's explanations and illustrations.

ELECTRICAL DISCHARGES, PLASMA, AND NETWORKS OF CURRENT IN SPACE

Benjamin Franklin (1706 – 1790) was one of the first natural philosophers of his day to determine that electricity existed in the sky. He invented the lightning rod that protected tall structures from being harmed by lightning electrical discharge. He was actually ridiculed by many religious people suspecting that God's anger could not be repelled by man's avarice. His 'pointed rod continued to work even saving the lives of bell ringers in church steeples. I consider him the very first pioneer to apply plasma technology and overcome the shunning by silly superstitions and dogmas.

James Clerk Maxwell (1831 – 1879) is famous for developing the relationship of electric currents and magnetic fields and how electromagnetic energy propagates through space. This idea developed into the electromagnetic spectrum and showed visible light to be only as extremely small part of the spectrum. This spectrum concept would further enhance the idea for both dark and glowing plasmas which are found throughout the celestial realm.

Kristian Birkeland (1867-1917), a professor in Norway, was the first person to propose that electric currents come from the Sun, flow into the upper atmosphere and cause auroras. These auroras were reproduced in the laboratory using his device called a 'terella'. He received hostility from British and American scientists of his day due to their prejudice against electrical effects in astronomical events.

Irving Langmuir (1881-1957), an experimenter of electrical discharges in low-pressure gases, discovered the double sheath or double layer (DL) effect in plasmas. This finding almost assures that electrical currents can travel in the so-called vacuum of space. He coined the term *plasma* because his experiments displayed the lifelike, self-organizing, self-sustaining behavior of these ionized clouds in the presence of electric currents and magnetic fields.

Hannes Alfven, another pioneer in plasma physics, predicted in 1963 the large-scale filamentary structure of the universe which contradicted the consensus of Big Bang cosmology by astrophysicists. He created the new fields of charged particle beams and magnetospheric physics winning him a Nobel Prize. He is still considered an outsider to consensus-astrophysics that relies more heavily on theoretical and mathematical means as opposed to experimental techniques.

Donald E. Scott, both an electrical engineer and astrophysicist, wrote *The Electric Sky* that challenges the many, as he calls them, myths of modern astronomy. He is part of the EU group who proposes that 1) electrical circuity drives the spiral arms of galaxies and the formation of stars; 2) Dark Matter, Dark Energy, neutron stars, and black holes are not real; 3) the stability of the solar system is chiefly caused by electromagnetic forces and not those of gravity; 4) the binary creation of planets and satellites is by the stress of electrical charge within the parent body; and 5) the fusion of elements is created in the extremely hot coronas of stars and not by the pressure of gravity at the center of stars. Amazingly I believe that Scott has the real sense of solar system formation and has 'nailed it'.

Nikola Tesla (1856 – 1943), an inventor and futurist, is best known for designs of the modern alternating-current electrical supply system. Tesla pursued ideas for wireless lighting and worldwide wireless electric power distribution in his high-voltage, high-frequency power experiments, but ran out of funding before he could complete it. His belief including my own is that the Earth is fed almost by unlimited electrical power from the Sun via the dark plasma of solar winds. He was trying to tap into this power through the use of worldwide grounding stations that would concentrate power from both the atmosphere and the ground. His hope was to transmit this power similarly as radio waves for communication. I propose that ancient technology already performed this miracle by using pyramidal structures.

What is needed now for all these valuable contributors of man's history and science is to connect these concepts into one congruent storyline. My book performs this function. I call it 'connecting all the dots' or completing the scientific jigsaw-puzzle pieces that have been created by man.

PREFACE

This storyline explains the natural reasons for worldwide catastrophes and one- after-another failed civilization. The book provides more detail and evidence for a periodic celestial marauder, Nemesis, as the follow-up to its introduction in the book, *The Great Deluge: Fact or Fiction?* Nemesis is the Sun's evil sister star that keeps returning to wreak havoc on the Sun's planets including Earth and its life forms. The Great Deluge or global flood was the final result of one of its more calamitous return visits 11,500 years ago. Other less damaging visits numbering three after the Great Deluge are explained.

Nemesis is an orbiting brown dwarf star that has not been detected yet by modern astronomy. Reasons for its undetectability are given as well as evidence of its distinctive destructive methods used on the planets of the solar system. Myths about its existence will be combined with scientific testimony. Petroglyphs or rock art and traditional archetypes are given as more confirmation by linking their pictorial representations to high energy plasma displays made in the laboratory.

The largest catastrophe occurred during and after the global flood explaining why mankind's technical progress 10,000 years ago was driven into a primitive Stone Age mode that leaves modern historians believing that humanity progressed from that time onward with small steps in technological improvements. Actually, proof is given that megalithic structures worldwide are older than the dating of the flood. The technology for making these wondrous structures is mysterious and forgotten and cannot be reproduced. The premise is that mankind had extremely advanced knowledge and technology which was wiped out leaving behind only shattered memories of a Golden Age that was then converted to myth and ancient traditions.

Should we still regard this mysterious Nemesis and its return once again? This book may answer that question for you.

III.

INTRODUCTION OF THE SUN AND ITS PERILOUS SISTER STAR

The Sun has a sister, dim, brown dwarf star that orbits the Sun about every 3600 years which is not well known or believed to exist. This periodicity may differ as much as several hundred years and is not perfect due to perturbations between it and close encounters with the Sun's planets. The orbit is highly elongated and passes very quickly through the inner solar system creating havoc with the Sun's and its own planets. Its suspected path comes from the southern hemisphere and is slightly inclined to the ecliptic plane of the Sun's planets. This dark brown dwarf can only be seen by special infrared telescopes and astronomers have not known where to look in the sky until very recently. Of course, the star's planets cannot be visually seen until close approach because they are not illuminated by either its own dim star or the Sun. The dwarf star and its planets are most of the time beyond the Kuiper Belt where sunlight is miniscule. The star more than likely passes through the region between Mars and Jupiter and creates debris mainly by electrical arcing and sputtering of its strong electromagnetic discharges between itself and other nearby planets, satellites, and asteroids.

IV.

POPULAR IDEAS ABOUT A NEMESIS STAR

This plausible star has been named Nemesis by the scientific community; the popular hypothesis is that Nemesis has an elongated elliptical orbit completely exterior to the Sun's planets and the Oort cloud which periodically comes close enough to disturb comets or minor planets either in the Oort cloud or possibly the Keiper Belt. Nemesis is either a red or brown dwarf star gravitationally connected to our Sun and its distance from the solar system center is in terms of light years.[1] The hypothesis was developed to explain a statistically determined cyclic period for 12 mass extinction events over the last 250 million years.[2] The Infrared Astronomical Satellite (IRAS) failed to detect Nemesis in the 1980s. The 2MASS astronomical survey failed to detect any nearby dim star in the 1990s. A more powerful infrared technology, the Wide-field Infrared Survey Explorer (WISE), is able to detect the more difficult brown dwarfs as cool as 150K out to a distance of 10 light years and released its final data in 2012.[3] No Nemesis was detected and presently NASA doubts that it exists. A final 2014 analysis of the WISE data has negated any Neptune sized object existing less than 700 AU away, and any Saturn sized object existing between the Oort cloud and out to ten thousand AU, and any Jupiter size object out to 1 light year or 63,000 AU assuming their surface temperature is above the detectable limit of about 150K. Admittedly, NASA was unable to detect any Kuiper belt objects with WISE since they fall below this temperature threshold.[4]

NASA officially stated that "recent scientific analysis no longer supports the idea that extinctions on Earth happen at regular, repeating intervals, and thus, the Nemesis hypothesis is no longer needed."[3] This statement only refers to claimed periodicities of 25 million or more years based on mass extinctions. Much smaller periods such as 3600 years are not even studied or even contemplated by NASA. This smaller period is claimed to exist by this paper and should be properly identified as the "Sitchin Hypothesis" naming the idea after its originator.

V.

NASA SKY SURVEY ISSUES

There are numerous issues with these NASA sky surveys in regard to looking for brown dwarfs that may be within much closer and unexpected orbital radius of 90 to 500 AU. The NASA mission is typically looking in terms of light years away for the Nemesis star and AU units within the solar system for asteroids. Recently discovered asteroids or comets could be easily confused with brown dwarfs. The lower part of the range of surface temperature between a typical asteroid like Vesta at 213K to 403K[5] is not much different from a cooled brown dwarf that can range from 750K to 2200K.[6] One ultra-cool brown dwarf, WISEPC J045853.90+643451.9 is reported[7]; how cool is it? Presumably, this dwarf is near the lowest temperature detectable of 150K. Also, interstellar extinction which is surrounding dust can reduce the stellar temperature and confuse its total luminosity. Some other reports in 2010 from WISE discussed two unambiguous brown dwarfs being detected as well as "candidates" that have unknown distances. Yes, distances are very difficult if not impossible to determine without available normal techniques. The normal techniques are parallax and astrometry which require sufficient light or heat emission combined with enough proper motion. Nemesis may provide neither enough heat especially if shrouded in dust or enough proper motion due to its highly-elongated orbit around the Sun. Another technique that cannot be applied is comparing the known luminosity or brightness to its observed brightness and computing the distance by the inverse square law. A good calibration of the known brightness for brown dwarfs does not exist. If the infrared surveys can actually track a suspected brown dwarf for a period of time and radar rebounding signals estimate its distance then orbital characteristics can be determined as is done for near-Earth asteroids. No such method of discovery has been reported for any brown dwarfs.

To locate good suspects a computerized program may select candidates from a list of objects with unusually higher proper motion (evidence of a star changing location over a short period of time) which can be precisely the wrong property for Nemesis. One NASA discussion states that fewer brown dwarfs are found in the cosmic neighborhood than previously thought; there exist as many as six luminous stars for every brown dwarf.[1] This claim is very suspicious being based strictly on the limit of detection. If nature's method of producing stars of all sizes including the smaller non-luminous ones and planets included, then the power law dictates that the majority of bodies produced must be the smallest ones and the largest bodies will be the rarest. Having an unknown albedo from the Sun's rays similar to an asteroid's reflected light any approaching dim dwarf star may be almost impossible to see in the normal visual range of telescopes. Indeed, brown dwarfs are as difficult as planets to discover in interstellar space unless somehow, they are observed to affect a nearby luminous object that can be readily observed.

This paper proposes that Nemesis passes much more closely and is more likely a very dim brown dwarf. Infrequently, but nevertheless present during the life of the solar system on many of Nemesis's passing orbits or crossings, it has more than likely lost one or more of its own planets and/or their satellites to the stronger gravity of the Sun's planetary system. Of course, direct collisions or near misses of each star's planets is also part of the storyline going backward in time hundreds, thousands, and millions of years. Collisions cause debris which can become sizable, threatening short and long period asteroids and comets that, in turn, increase the prospect of future lethal collisions with the inner planets.

WHY CANNOT THE NEMESIS BROWN DWARF STAR BE FOUND?

This elusive celestial body may vary in size and mass from that of one of the outer planets in the lower limit to the upper limit of 2 to over 300 times the mass of Jupiter for brown dwarfs. Below 13 Jupiter masses (M_j) the star is called a sub-brown dwarf or a Y-dwarf due to its theoretical energy output. Brown dwarfs were theorized in the 1960's and verified discoveries were made in 1995. Teide 1 was the first free-floating at 57 ± 15 M_j having a 3.78 radius of Jupiter. Since Teide 1's discovery at 400 light years away, more than 1,800 brown dwarfs have been discovered. Gliese 229B, a companion brown dwarf to a Gliese 229A - a red dwarf, was confirmed that same year at only 19 light years away. This brown dwarf was measured as 21 to 52 M_j or 0.02 to 0.05 solar masses.

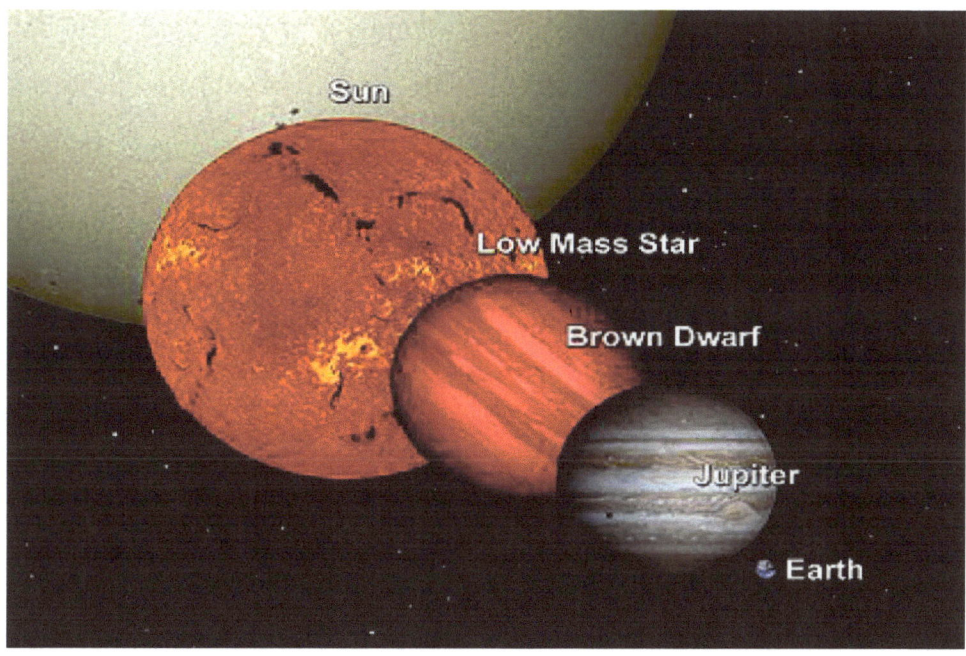

Figure 1: Comparative Size of Brown Dwarfs and Other Celestial Objects

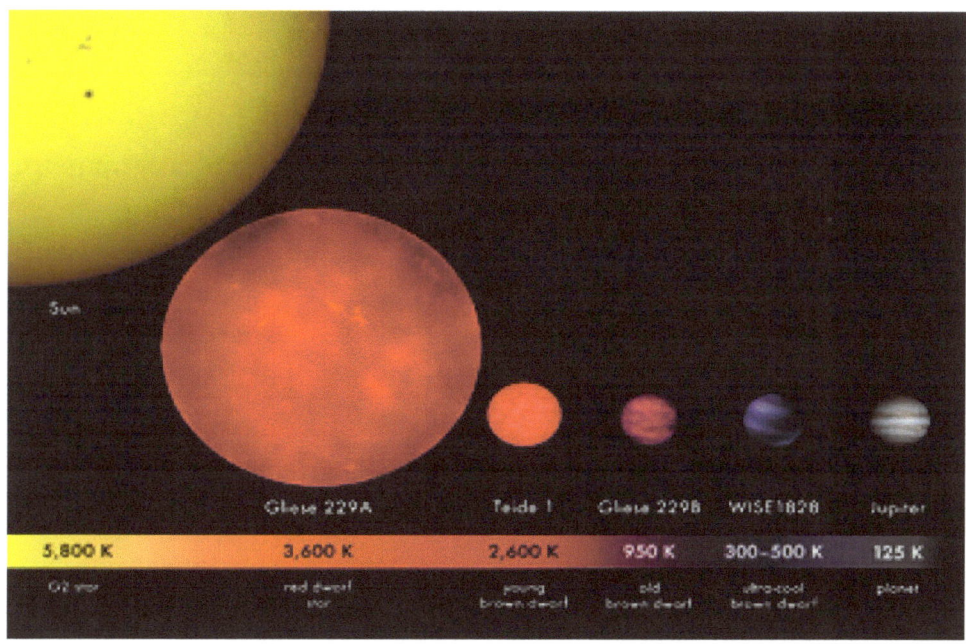

Figure 2: Comparative Sizes and Temperatures of Different Brown Dwarfs
Shown Beside the Sun and Jupiter
(The Newly Discovered Y-Dwarfs Have Temperatures of About 225 K.)

The near infrared spectrum of these brown dwarfs clearly exhibits methane absorption previously only observed in the atmosphere of the gas giant planets. A connection or similar genesis for these celestial bodies is suspected. Recently discovered 12 light years from the Sun are Epsilon Indi Ba and Bb, a pair of brown dwarfs bound gravitationally to a Sun-like star. In August of 2011 NASA's Wide-field Infrared Survey Explorer (*WISE*) discovered six Y-dwarfs with temperatures as cool as the human body or 298 K (25° C/77° F). The coolest at 7.2 light years away was verified in April, 2014, as 225 K to 260 K.

In 1999 some brown dwarfs were shown to be X-ray sources having magnetic fields. Their coronas are hypothesized to cease existing below 2800 K and become electrically neutral. Perhaps the stars' magneto-sheath shields the electromagnetic spectrum except for some of the infrared. In 2013 Hubble and Spitzer space telescopes showed brown dwarf weather of wind driven, planet-sized clouds. This depiction may be the result of orbiting planets inside a bloated corona sheath. In 2015 the first terrestrial-mass planet was found orbiting a brown dwarf.

In the search for exo-solar planets by NASA, numerous brown dwarfs were found orbiting larger yellow dwarfs such as the Sun. Enough data was collected and studied statistically to show brown dwarfs that orbit within 3 to 5 AU are less than 1 % of the stars with a mass similar to the Sun. This desert region of brown dwarfs is a mystery understandably when related to the nebula hypothesis and accretion disk paradigms. Quite possibly, the electric and magnetic fields generated by both stars create a balance of repulsive verses attractive forces to maintain this average gap of 3 to 5 AU - if they are either formed together or captured. The brown dwarf postulated by this paper does not go any closer than 3 AU as it passes twice through the Sun's ecliptic plane near the Main Belt of asteroids for each orbit. The Sun's sister brown dwarf is postulated to have been captured somewhat later in the formation of the Sun due to its proposed highly inclined and elongated orbit.

Another glimpse of our interstellar neighborhood reveals the incredibly far-reaching forces between stars. A red dwarf and known closest star to our Sun is Proxima Centari which orbits the Centari A and B binary system with a separation of 12,950 AU and an orbital period of about 550,000 years. This study tells astronomers that the Nemesis orbital period around our Sun is only a small fraction of all possible orbits that include Proxima Centari's orbital size.

Given the above information collected by NASA since the 1990's the following summary is made:

1. Nemesis could either be a large planet with its own satellites or a brown dwarf with its own planets.

2. If planetary-like, then the size of Nemesis could be comparable to Jupiter in mass and radius. Jupiter's surface temperature ranges from 165 K to 112 K near the

threshold of detectability of 150 K. Given Jupiter's angular diameter of 29.8" to 50.1", the expected angular diameter at 100 AU instead of 5 AU will be about 2". There would be no detectable light caused by the Sun for this angular diameter.

3. If Y-dwarf-like, then the size of Nemesis would range from 2 to 13 Mj; the coolest at 3 to 10 Mj were measured at 225K to 260 K which is not that far above the detectable temperature limit.

4. If like a typical brown dwarf, then the size of Nemesis would range about 25 to 300 Jupiter masses with typical parameters such as of Teide 1 of 55 Mj, 2600 K surface temperature (based on theoretical calculations), and 3.78 times Jupiter's radius.

5. Nearest known brown dwarf, Luhman 16, is 6.5 light years; Kuiper Belt Objects (*KBO's*) are found roughly between 30 and 55 AU with some as far as 60 AU from the Sun. The true meaning of this simple tabulation is that nothing is known to exist presently from the Sun between 60 AU and 6.5 light years x 63,000 AU = 410,000 AU with the exception of some Scattered Disk Objects (*SDO's*). Has NASA honestly and sufficiently searched this vast volume of space surrounding the Sun?

6. Brown dwarfs are known to exist by themselves or have companion dwarfs or companion terrestrial-size planets with orbits a few AU or smaller.

7. Brown dwarfs are known to orbit Sun-like stars.

8. Brown dwarfs are known to display electromagnetic (*EM*) characteristics such as flaring, changing magnetic fields, and auroras; however, only radio emissions are received during flaring episodes.

9. Brown dwarfs are not very luminous at visible wavelengths. Many are known to have surrounding disks of dust and gases. Possibly, the disk is actually more spherical with a magnetosheath that shields radio and light emissions. The surface temperature of its expanded corona boundary or magnetosheath for many brown dwarfs could be less than the 150 K detectable limit. This lower sheath temperature could be easily lowered and maintained by a brown dwarf's continued periodic crossing of another magnetosheath such as our Sun.

Figure 3: Artist's Impression of Dust and Gases Surrounding a Brown Dwarf

(More likely, the surrounding materials could be more spherical in shape.)

All this recent evidence collected over the past two decades leads most reasonable thinkers to believe in the plausibility of a Nemesis intruder – either a large planet or brown dwarf that is hypothesized in this paper. Is there any evidence of known celestial bodies that would behave similarly as the postulated Nemesis with its very inclined, very elongated orbit? Yes.

A scattered disk object, discovered in 2003, was found to orbit the Sun every 11,400 years in an incline to the ecliptic of 12 degrees. The perihelion is 76 AU with the projected aphelion at 937 AU. Sedna, also considered a minor planet, is listed as having a temperature of 12 K; obviously, a light detecting device instead of an infrared detector made the discovery. This minor planet's genesis is a mystery, but the author's guess is that there is a direct connection to Nemesis' trajectory around the Sun.

Figure 4: Comparative Sizes of Minor Planets Near and Beyond Neptune

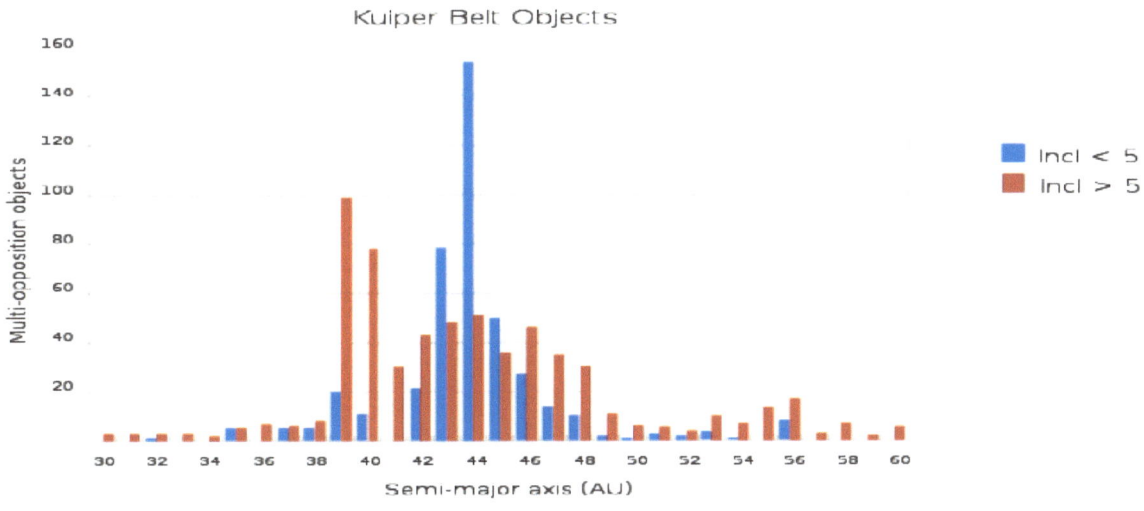

Figure 5: A NASA Generated Chart of Various Populations of KBO's and Their Locations and Inclination

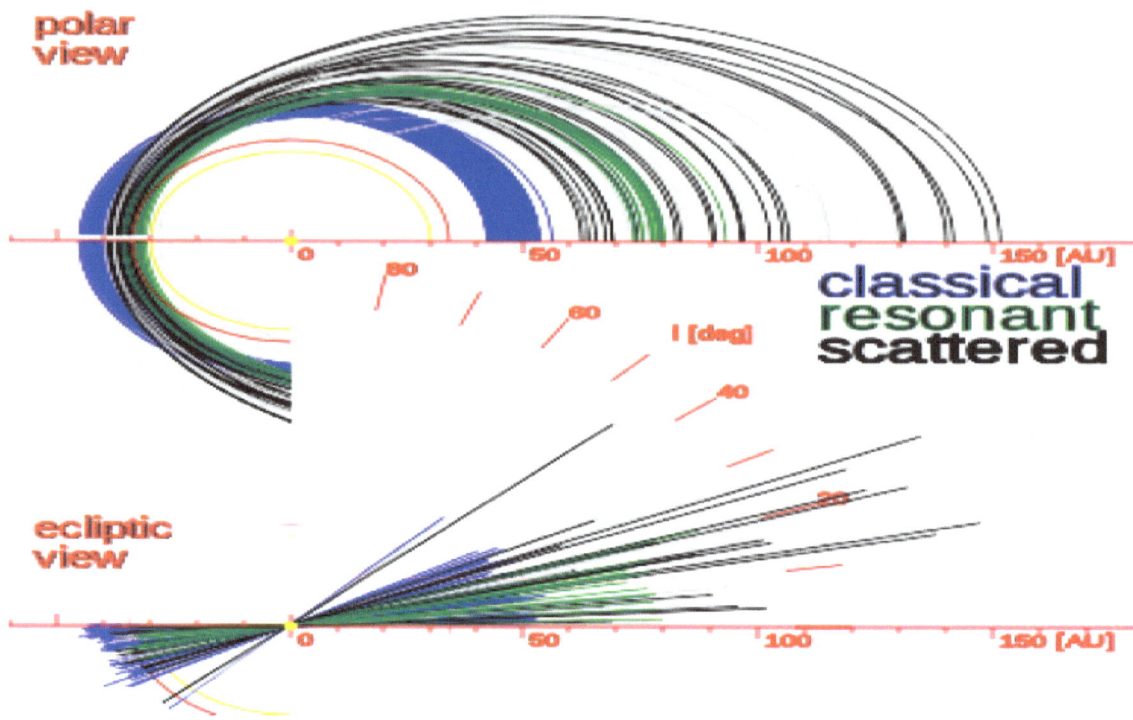

Figure 6: Representation of Solar System's Asteroids Including the Scattered Disk Objects (SDO's) Which Are Asteroids in Highly Elongated and Inclined Orbits

Nemesis is postulated to have an inclined orbit of some of the highest inclinations of SDOs of about 20 to 30 degrees which may take it beyond the expected torus of investigation for NASA astronomers. The apoapsis is projected to be more than 400 AU with a semi-latus rectum of 3 AU which creates a high ratio for the major to the minor axes of its elliptical path. This very narrow ellipse eliminates the chance for any appreciable proper motion being detected across the sky with respect to Earth observers. A computerized algorithm for detecting proper motion of this star is almost impossible.

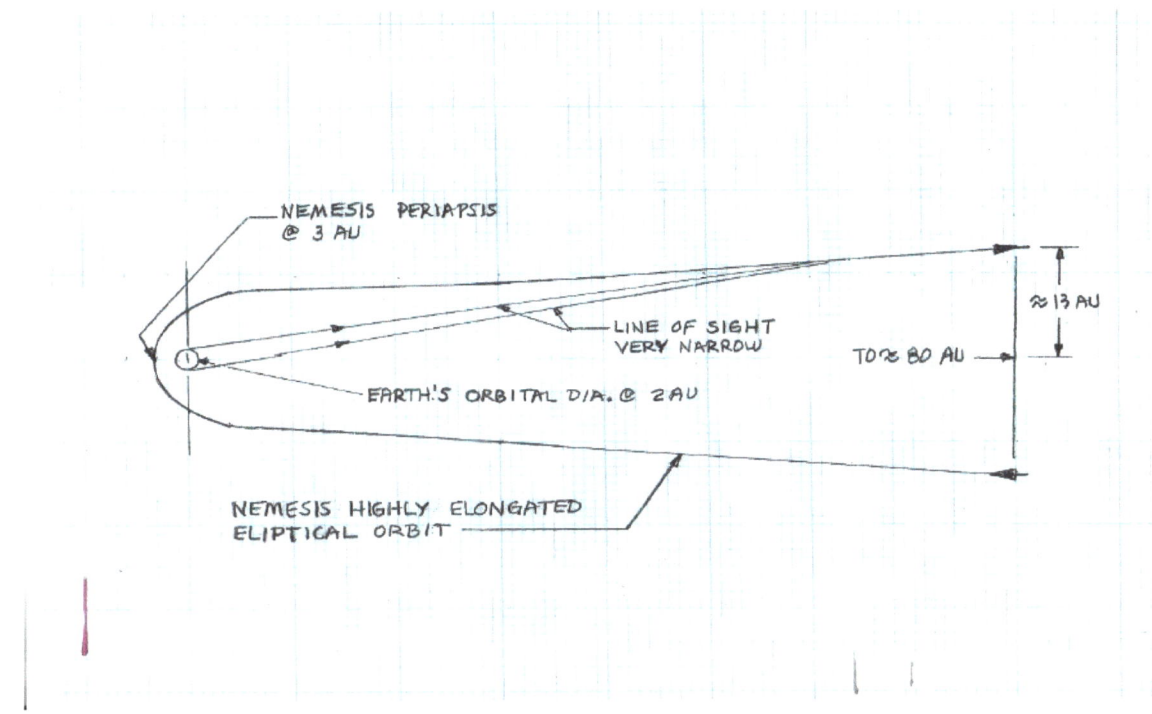

Figure 7: Diagramming the Difficulty of Applying 'Proper Star Motion'
and/or Parallax
(Nemesis' Orbit is an Extremely Elongated and Inclined Ellipse.)

Optical interferometers using a large array of telescopes on Earth can study objects at unprecedented detail such as the surfaces and diameters of stars. They have a spatial resolution of an amazing 4 milliarcseconds. Why not use these optical interferometers to search for Nemesis? Limited aperture area and atmospheric turbulence limits ground-based arrays of interferometers to observations of comparatively bright objects. Hence, the dimmest objects such as Nemesis do not have enough light; this limitation is called the "thinned-array curse".

The NEOWISE project took photographs of 600 near-Earth objects (NEO's) within 2 AU of the Sun because astronomers were able to combine visible light observations with infrared heat measured by the WISE spacecraft. The Nemesis-type dim star or planet cannot be found this way because no measurable reflected light from the Sun is presently available.

NASA's Wide-field Infrared Survey Explorer (WISE) has discovered 100 new brown dwarfs, but astronomers have not yet examined all the immense quantity of data. Almost ¾ of a billion objects (asteroids, stars, and galaxies) have been photographed. Infrared wavelengths have been scanned from January 2010 to February 2011 for the entire sky about 2.0 times. A six-month gap between scans enables astronomers to compare the surveys for moving objects. When candidates are selected due to their motions NASA's Spitzer Space Telescope helps to confirm and narrow the list using such methods as parallax. Telescopes on Earth are then able through spectrometry to identify expected molecular signatures of water, methane, and ammonia for many of the brown dwarfs. Objects that meet the parameters of brown dwarfs are selected with the help of devices called CCDs that can automatically choose objects with certain infrared intensity and proper motion with respect to the observer. The arduous task of comparing photographic plates is no longer necessary. For certain distances from the Sun a certain combination of these parameters is expected. The visual clue and algorithm are that over time the faster object is considered closer and the slower object is considered farther away similar to observing planes flying at different altitudes. However, a celestial object on an extremely elongated orbit such as the Nemesis system will seem to be much farther away because the proper motion is seen as much less than expected. See the previous diagram for the dilemma of not having enough proper motion.

Photometry is an astronomical method of determining the distance of various celestial objects from Earth. The object's apparent brightness magnitude is compared with its known absolute brightness. Then the inverse-square law is applied that defines how energy flux decreases with distance. However, for the recent discovery of brown dwarfs with all their variations in theories of energy output and possible shielding by dust and gases or by a special expanded corona, the absolute magnitude for brown dwarfs is still a mystery.

A summary of the major difficulties for NASA scientists finding Nemesis are:

1. Extreme dimness allows only the best performing infrared telescopes to be used.

2. The outer envelope of certain brown dwarfs may be too close or below the detectable limit of 150 K.

3. Many brown dwarfs have surrounding disks of dust and gases that may shield other electromagnetic signatures from being emitted. X-rays are only emitted when certain brown dwarfs begin to flare indicating that an envelope if not broken does shield EM signals.

4. The possibly very short proper motion due to Nemesis' elongated trajectory may either make it impossible to be selected as a candidate, or astronomers confuse its location as being much farther away.

5. Photometry, the process of using the inverse-square law, cannot be applied with any accuracy for brown dwarfs.

6. The enormous amount of data has not all been analyzed fully or properly.

7. The paradigm of discounting brown dwarfs just outside the Kuiper Belt can cause certain data to be labeled anomalous and mysterious. Such data may be shelved for later analysis.

Dramatically, NASA's online website headlined, "NASA's WISE Survey Finds Thousands of New Stars, But No Planet X", in March, 2014. This hypothesized planet also dubbed Nemesis or Tyche was reported to not exist beyond the orbit of Pluto. The following diagram was produced to show the WISE detectable region where no bodies larger than the outer planets exists. This unlikely conclusion was given by Kevin Luhman of the Center for Exoplanets and Habitable Worlds at Penn State University in Pennsylvania. However, conflicting statements are made. The 'Planet X' search was based on extinction events millions of years apart and also on irregular asteroid and comet orbits. NASA admits that each new study of the data reveals celestial bodies missed in previous studies. And, in 2013, WISE was reactivated to look for potentially hazardous near-Earth objects (*NEO's*). Very recently in August 2015 the Jet Propulsion Laboratory (JPL) released a news item: "Tracking A Mysterious Group of Asteroid Outcasts". Much interest developed about finding the source of the Euphrosyne family of dark asteroids, KBO's, on highly incline orbits in the outer asteroid belt. Apparently, for NASA and JPL the interest and/or concern for Planet X or Nemesis has **not** gone away.

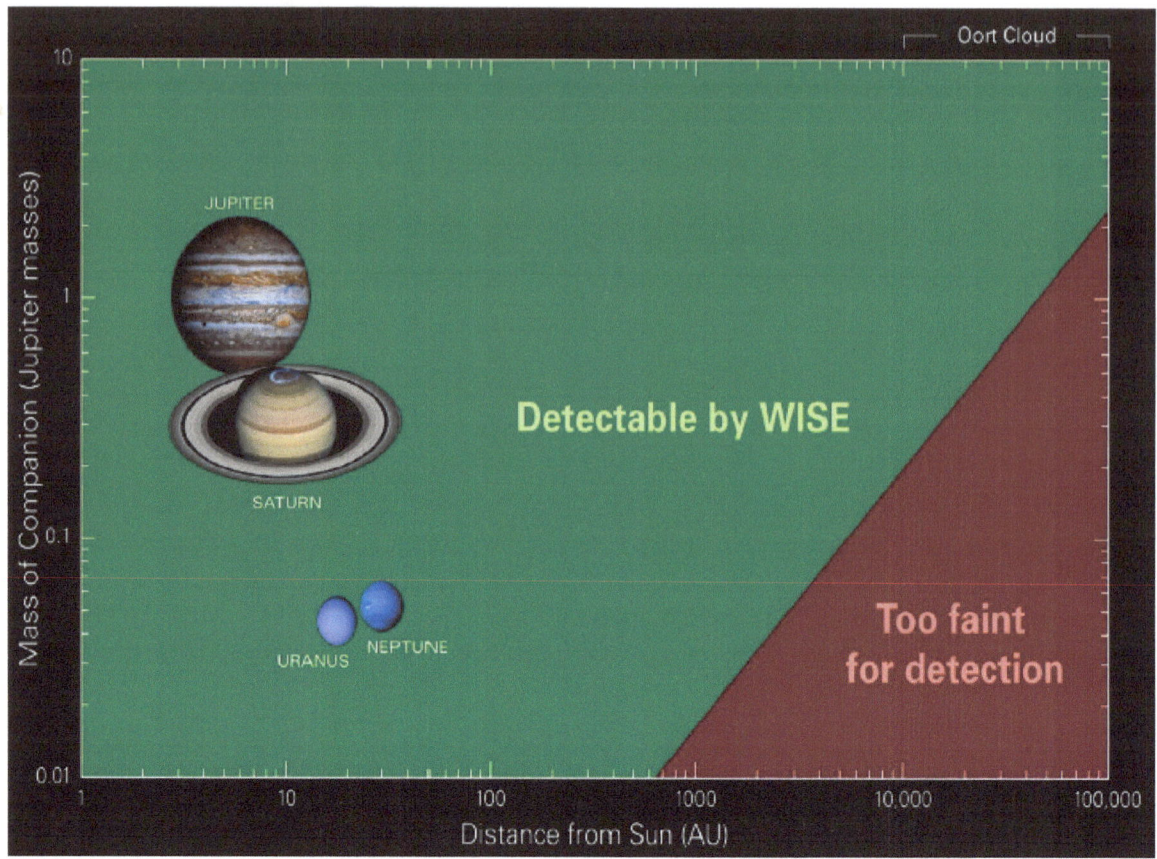

Figure 8: NASA's Representation of the Detectable Region for WISE Infrared Sky Survey

(This diagram assumes that no celestial body is near or below the threshold of 150 K detectability and closer bodies have enough proper motion to be selected for analysis.)

What if Nemesis has already been discovered and its information is classified? Would the governmental authorities who rule over NASA worry about the release of such information to the public? Would the public become so shocked and unstable knowing that impending global disasters on Earth may occur in 10's or 100's of years. Some conspiracy theorists even conjectured that parts of the Google southern sky are blocked out due to NASA orders. These possibilities may exist due more to this author's suspicions coming from another source, the study of forensics in the solar system.

NASA and ESA space probes to all the planets and most of their satellites, to comets and asteroids, and to the magneto-sheath of the Sun reveal a very distinct reality that is currently denied. The majority of surface features on these objects indicate the result of electromagnetic phenomena that overshadows the conventional idea of asteroid collisions.

These surface features, as promoted by an independent group of scientists associated with the Electric Universe.com prove that very energetic plasma arc discharges have occurred. The configurations of these features, especially those well photographed on Mars, reveal precisely what can be duplicated in the study of plasma in laboratories on Earth. NASA and JPL have remained strangely silent on this matter. The Electric Universe people simply accuse consensus science as being caught in the paradigm of the impossibility of lightning-type electric currents or plasma passing in a vacuum of space between closely passing celestial bodies inside the solar system. And, of course, for NASA the only close encounters are between planets and asteroids. But, the evidence for electrical and magnetic forces creating and maintaining the stability of the solar system keeps mounting. The studies of changing solar winds; witnessed electrical discharges on comets, on the Moon, and on Jupiter's Io; planetary auroras correlated to solar winds; and the magneto-sheaths of the Sun and planets all point to EM phenomena ruling interstellar and interplanetary environments.

The major question for this paper is what has caused these countless disruptions of planetary and satellite surfaces by electrical means. Postulating that planets and stars carry a varying unbalance of electrical charge, their close encounters over millions of years will cause lightning bolts between the two passing objects just as happens between Earth's surface and clouds. And, how do close encounters occur? The Nemesis hypothesis creates these close encounters on an approximate periodic basis over millions of years if its orbit carries itself and companion planets through the Sun's own planetary system. As part of the hypothesis, the periodic disruptions will result in a myriad of results for each period due to different planetary locations and different exchanges of electrical charge. Hopefully, NASA and JPL will begin a dialogue with the Electric Universe scientists in the near future.

The most important criticism of this paper's Nemesis hypothesis is the plausibility of orbital mechanics proving that such an orbit is possible with the given gravitational forces that certainly cannot be denied. Furthermore, can this orbital combination of the two stars with their planets remain stable over thousands or over millions of years? Consideration for orbital mechanics will now be addressed.

VI.

A SANITY CHECK USING KEPLER'S LAW AND ORBITAL MECHANICS

Orbital mechanics applies practical problems of the motion of rockets and spacecraft including natural astronomical bodies such as star systems, planets, moons, and comets. Their motions are calculated from Newton's laws of motion and Newton's law of universal gravitation. General relativity is a more exact theory, but is only necessary for greater accuracy or in high-gravity situations such as orbits close to the Sun and near-Earth orbits of spacecraft and artificial satellites. Johannes Kepler published a model for planetary orbits in 1605 and Newton's more general laws were published in 1687. Kepler's Laws can also be applied to the simple case as the Nemesis star or planet orbiting the Sun without the complications of other companion planets or moons.

Let's check with Kepler briefly and use his Third Law to make some simple calculations. Is an orbit of 3600 years around our Sun of its brown dwarf companion a reasonable assumption? The masses of brown dwarfs recently discovered as companions in other star systems, such as Teide 1 and Gliese 229B, are determined to have less than 0.052 solar masses (M_Θ) and equal to or less than 0.076 M_Θ (80 M_J) respectively.[8,9] An approximate orbital radius for these samples can be determined by Kepler's Third Law if it is assumed that the body orbits the Sun. The equation is:

$$G \times (m + M) \times t^2 = 4\pi^2 \times r^3$$

where 'm' is the mass of Nemesis and 'M' is the mass of the Sun. The mass of the Sun completely overwhelms the smaller masses of either an outer planetary mass such as Jupiter or a brown dwarf mass such as the chosen typical Teide 1's mass.

The examples chosen for Nemesis are a planet that is 2 times the size of Jupiter (0.00192 x mass of Sun) and a brown dwarf the size of Teide 1 (0.0524 x mass of Sun). Hence, the gravitational parameters of u = G (M_{sun} + 2m_j) = G (1.00192 x M_{sun}); and, u = G (M_{sun} + m of Teide1) = G (1.0544 x M_{sun}) are hardly distinguishable from each other.

The simplified Kepler's Third Law equation using periods and orbital radii is applied –

$$P_1^2 / P_2^2 = R_1^3 / R_2^3$$

where the Nemesis period of P_2 = 3600 years is utilized and R_2, the approximate radius of Nemesis's orbit is 234 AU assuming a round orbit. The other orbital radii would be 246.7 AU for a typical brown dwarf and 234.4 for a typical planet the size of an outer planet.

The above calculation does not consider the elongated ellipse that is postulated. For this approach, the latus rectum = L of its elliptical orbital path is used as a datum point. That suspected datum point is where the Nemesis system crosses the ecliptic plane of the solar system planets near the Main Belt of asteroids between Mars and Jupiter. This distance is chosen at 3 AU although the average distance of Main Belt asteroids is 2.7 AU. The crossing of Nemesis at these two points on its inclined orbit is thought to continue for millions of years and keep the Main Belt in existence; this orbital possibility will be called Version #1. Conventional thinking is that nearby Jupiter caused gravitational resonances and prevented these asteroids from either dispersing or combining together over millions of years. Consensus science also believes that the asteroids were originally part of the residual accretion of the primordial disk of dust that rotated around the proto-sun. However, the space probes have proven that these chunks of matter look much like debris caused by collisions or high energy plasma arc sputtering of existing hardened celestial bodies.

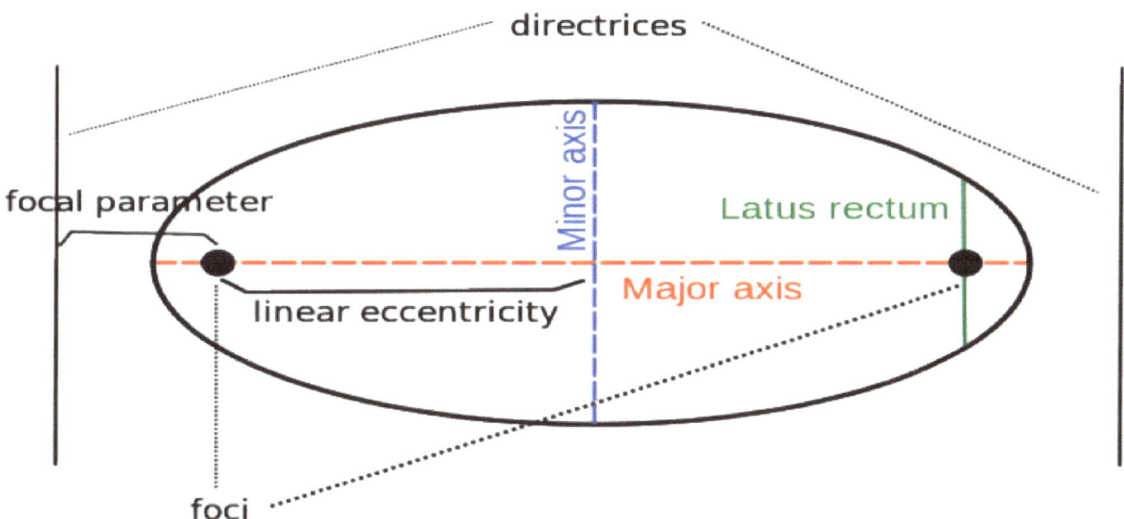

Figure 9: Ellipse parameters showing major and minor axes and the latus rectum

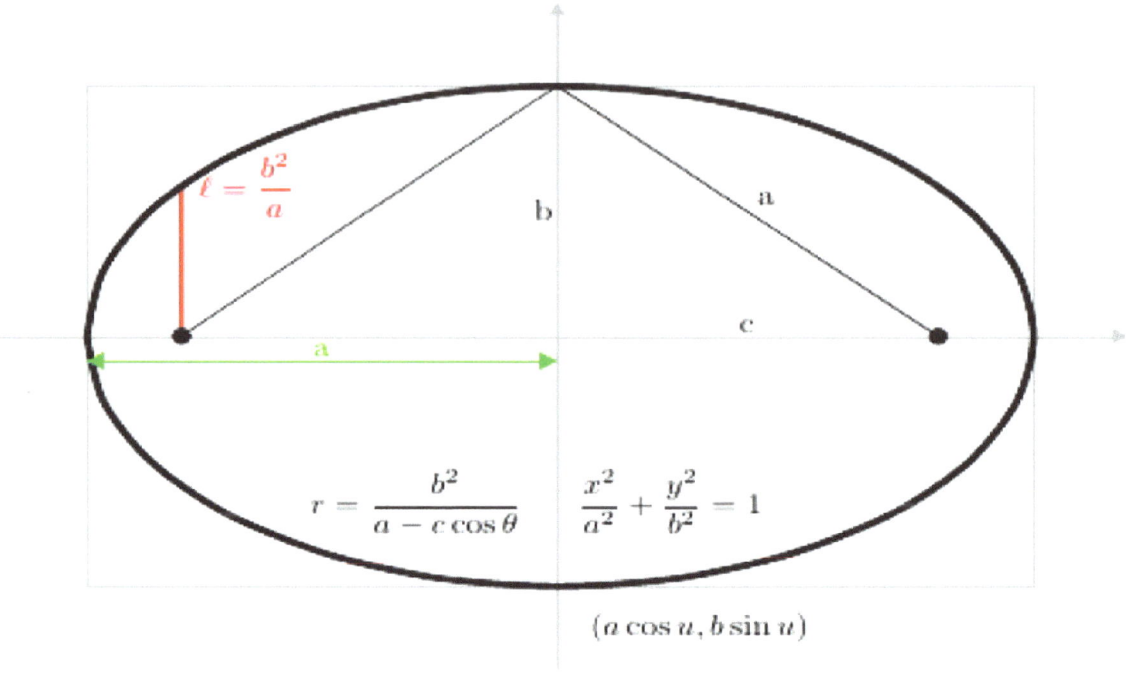

Figure 10: Standard forms of an ellipse from internet's online topic of Conic Sections

Other possibilities or versions for orbital trajectories are that of Version #2 - the periapsis of Nemesis is 3 AU thereby hovering over and disturbing the Main Belt during each periapsis of its orbit; and Version #3 - the periapsis is about 30 AU from the Sun and very close to Neptune's orbit but well outside the ecliptic plane due to its inclined orbit. **Let's now test the Nemesis hypothesis to check in realistic terms the orbital mechanics?**

The calculations for these orbital paths begin with the given orbital period of 3600 years and the gravitational parameter of –

$$u = G \left(M_{sun} + m_{Nemesis} \right)$$

where G is the gravitational constant = 6.673 x 1011 m³/kg x s²; M_{sun} = mass of Sun = 1.989 x 10³⁰ kg; and $m_{Nemesis}$ = mass of typical brown dwarf (Teide 1 is chosen, which is 0.052 times the mass of the Sun.

Hence, the gravitational parameter in this case is –

$$u = 13.963 \times 10^{19} \ m^3/s^2$$

For a larger brown dwarf such as Gleise B the gravitational parameter is –

$$u = 14.281 \times 10^{19} \text{ m}^3/\text{s}^2$$

For a typical outer planet such as Uranus being Nemesis the gravitational parameter is –

$$u = 13.273 \times 10^{19} \text{ m}^3/\text{s}^2$$

As is indicated above, the mass of Nemesis, if it is either as massive as a brown dwarf or as an outer planet, does not affect its orbital path as much as the Sun's mass and the predicted orbital period of Nemesis.

Utilizing the equation for the orbital period for an elliptical path will reveal the major axis, 2a –

$$\text{Orbital Period} = T = 2\pi \times (a^3/u)^{1/2}$$
$$a = 3.57 \times 10^{13} \text{ m} = 238 \text{ AU, and}$$
$$2a = 476 \text{ AU} = \text{major axis}$$

Letting L, the semi-latus rectum, equal 3 AU and the semi-major axis = 238 AU for Version #1, the following operations are performed –

$$L = b^2/a, \text{ and}$$
$$b = \text{semi-minor axis} = 26.7 \text{ AU};$$
$$c = (a^2 - b^2)^{1/2}, \text{ and}$$
$$c = \text{linear eccentricity} = 236 \text{ AU};$$
$$r_p = \text{periapsis} = (a - c)$$
$$r_p = 2 \text{ AU}$$

which means that Nemesis comes close to the Martian orbit and Main Belt of asteroids but never approaches the outer planets.

In the case of Version #2, let the periapsis, r_p = 3 AU and a = 238 AU, then the following operations are performed –

$$c = a - r_p = 235 \text{ AU, and}$$
$$b = (a^2 - c^2)^{1/2} = 37.7 \text{ AU, and finally}$$
$$L = \text{semi-latus rectum} = b^2/a = 6 \text{ AU}$$

which means that Nemesis comes close to Jupiter with an orbital radius of 5.2 AU, but is away from the ecliptic plane of Sun's planets.

For Version #3, let the periapsis, r_p = 30 AU, making the resulting apoapsis, r_a = 476 – 30 = 446 AU. The following operations are performed –

$$c = a - r_p = 208 \text{ AU} = \text{linear eccentricity, and}$$

$$b = (a^2 - c^2)^{1/2} = 115 \text{ AU;}$$

$$L = b^2/a = 55.6 \text{ AU}$$

which means Nemesis stays well outside the orbits of the Sun's planets. Its effect on the solar system would be possible perturbations of its own planets or satellites, the consequences of barycenter changes, and the electrical field effects of the Sun and Nemesis passing through each other's magneto-sheaths. This version provides the most possible stability of the system over much longer periods of time such as millions of years. A table and diagrams shown approximately to scale are provided to summarize the orbital characteristics of these three plausible orbital modes.

Table 11: Orbital Characteristics in AU Units for Different Predicted Paths of Nemesis

(These paths are shown in the following diagrams.)

Version	Semi-major axis (a)	Semi-minor axis (b)	Linear ecc. (c)	Semi-latus rectum (L)	Periapsis (r_p)
#1	238	26.7	236	3	2
#2	238	37.7	235	6	3
#3	238	115	208	55.6	30

Figure 12: Three Scaled Diagrams of Proposed Orbits of Nemesis

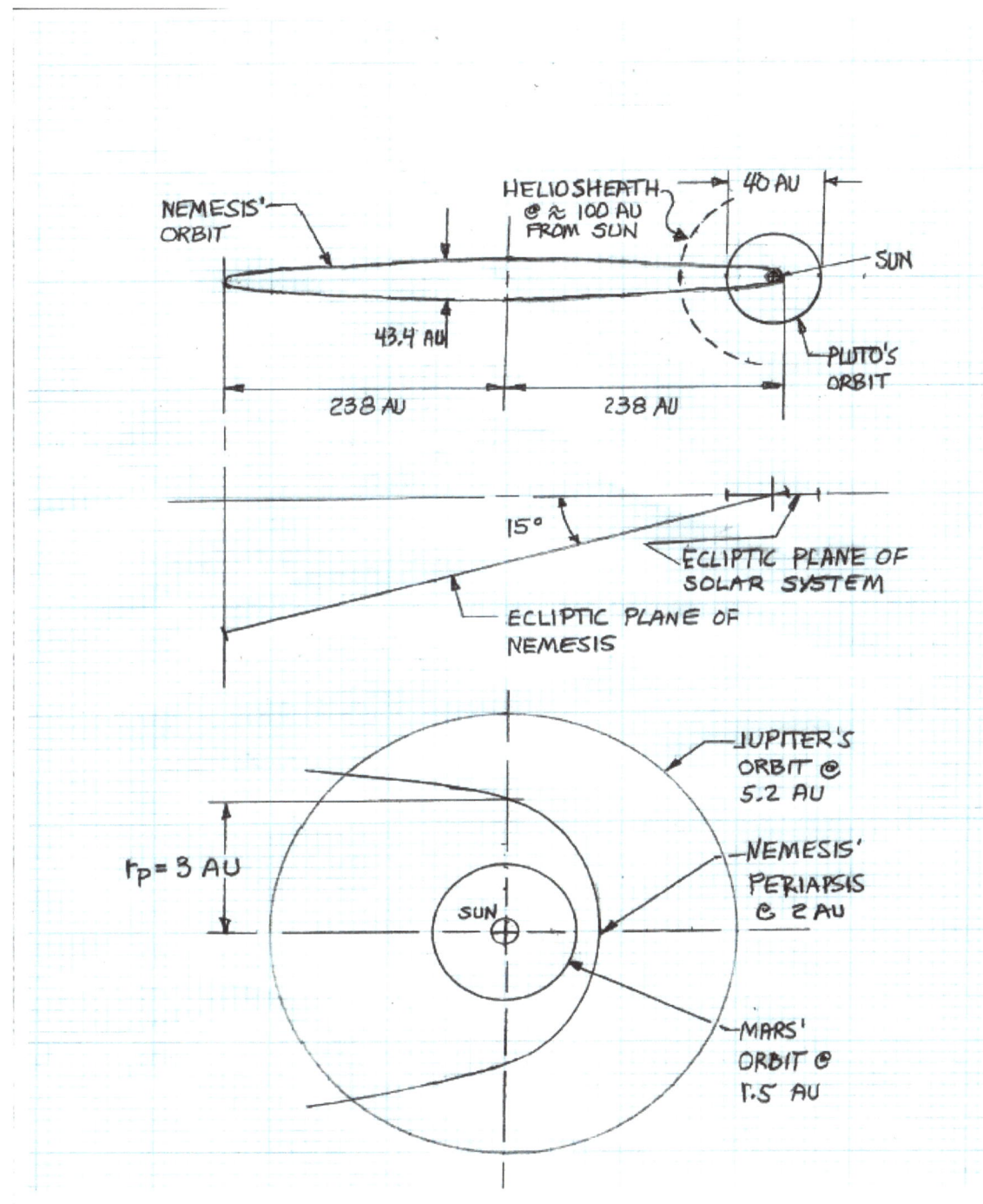

Version #1: Diagram to Scale of Nemesis' Orbital Path with a Periapsis of 2 AU

Figure 12: Three Scaled Diagrams of Proposed Orbits of Nemesis

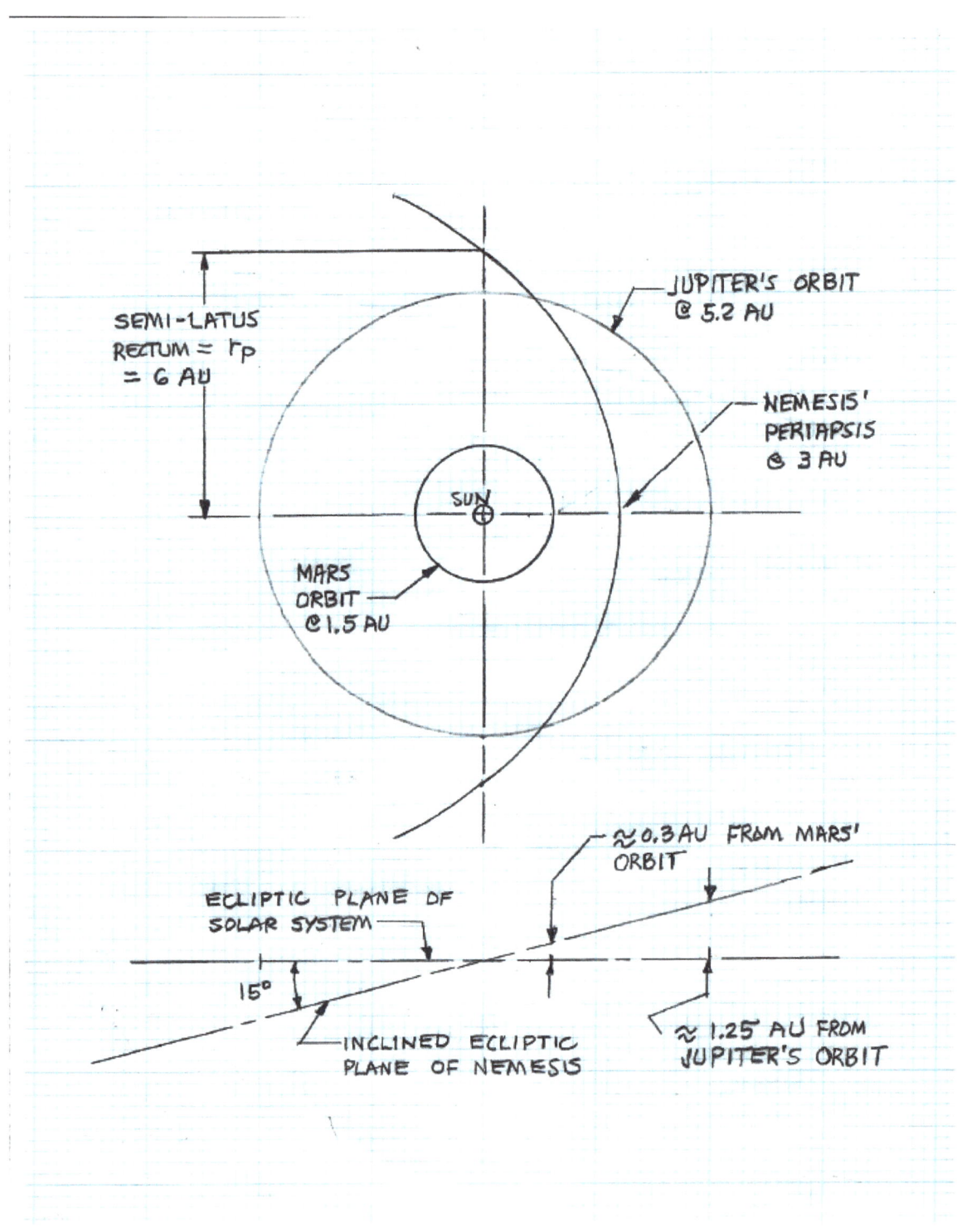

Version #2: Diagram to Scale of Nemesis' Orbital Path with a Periapsis of 3 AU

Figure 12: Three Scaled Diagrams of Proposed Orbits of Nemesis

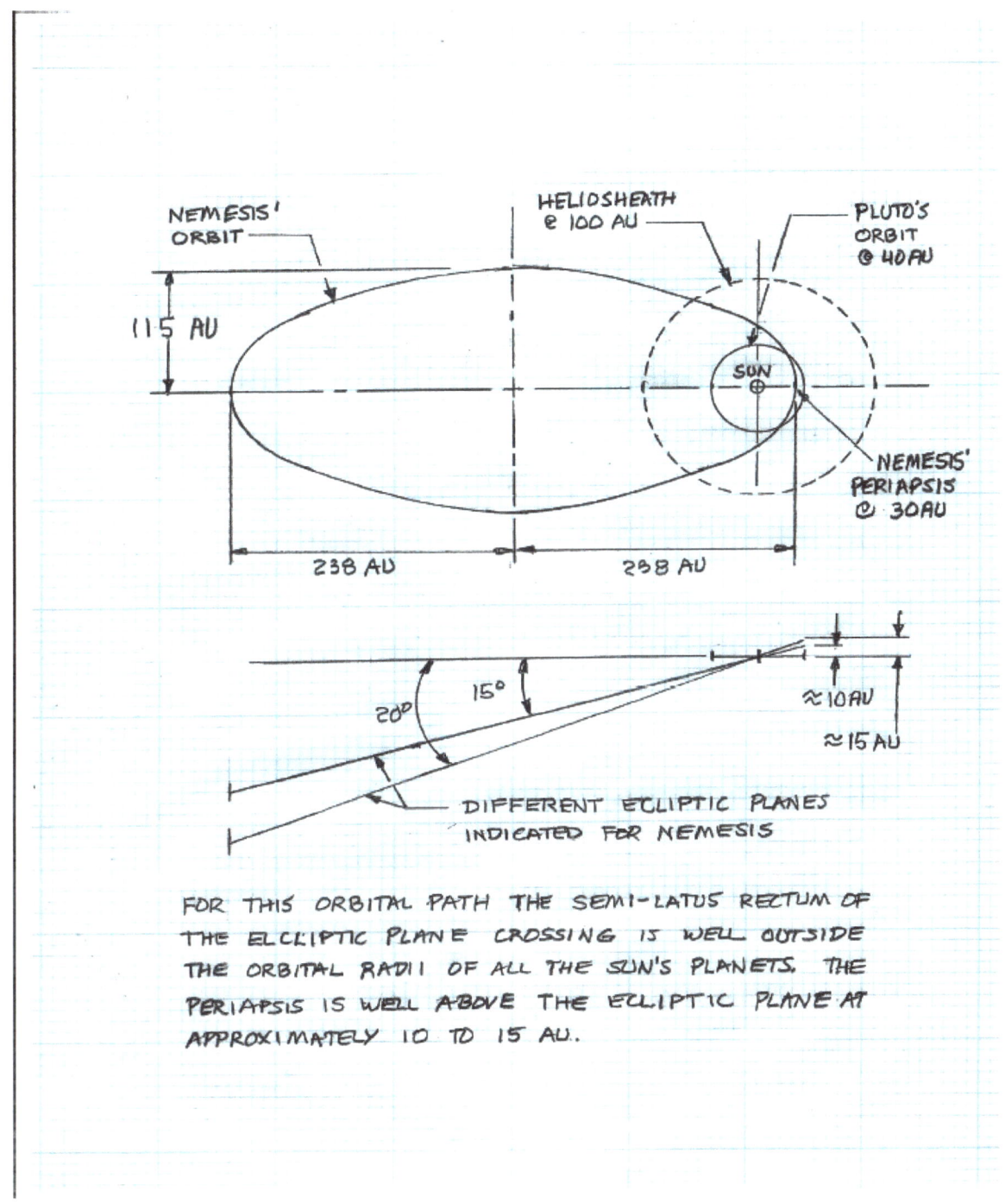

FOR THIS ORBITAL PATH THE SEMI-LATUS RECTUM OF THE ELCLIPTIC PLANE CROSSING IS WELL OUTSIDE THE ORBITAL RADII OF ALL THE SUN'S PLANETS. THE PERIAPSIS IS WELL ABOVE THE ECLIPTIC PLANE AT APPROXIMATELY 10 TO 15 AU.

Version #3: Diagram to Scale of Nemesis Orbital Path with a Periapsis of 30 AU

The question still remains whether any of these overlapping celestial systems are stable over thousands and millions of years. Certainly, the closest encounters and multi-conjunctions or alignments of several planets and a star will lead eventually to wild perturbations over large intervals of time with repeated orbital crossings. Computerized simulations of these Nemesis' orbitals paths based strictly on Newton's laws of motion and his universal gravitation equation will certainly indicate natural instability. But, are these simulations interpreting nature properly? Electromagnetic factors should also be utilized.

The recent discovery of two very unusual KBOs or dwarf planets does lend more credibility to this paper's Nemesis Hypothesis. These minor planets are also termed Trans-Neptunian Objects (TNOs) since their orbits are a greater average distance than Neptune's orbit. In fact, their orbits are significantly larger and more eccentric. Their data is taken from Wikipedia and tabulated below.

Table 13: Unusual Orbital Parameters for Some Trans-Neptunian Objects (TNOs)

TNO Designation	Discovery Date	Orbital Period in Years	Aphelion in AU	Perihelion in AU	Inclination
90377 Sedna	Nov, 2003	11,400	937	76	11.9°
2012 VP113	Mar, 2014	4452	461	80	24.0°

Their orbital characteristics are reminiscent of Nemesis' predicted features. Much speculation is occurring about why these TNOs have such orbits. Sedna is not near Neptune's orbit and could not be possibly perturbed by this planet. Academics are making some concession that it was captured from another star system when the Sun was still inside an open star cluster. The capture of free exo-solar planets after an open star cluster disperses is currently not an option by academics and their celestial mechanics. Speculation for 2012 VP113 is similar except a regular TNO planet of several Earth masses is also suspected. A possibility not known to be considered by NASA is that a Nemesis-like star can also perturb

2012 VP into this orbit as well with other scattered disk objects (SDOs) over immense periods of time with small cycles of 3600 years.

A curious reporting by NASA via Wikipedia is that all major bodies with a semi-major axis of over 150 AU and perihelia greater than Neptune's have arguments of perihelion clustered near 340 degrees which begs for a similar formation mechanism. Could this mechanism be the result of perturbations and/or close encounters of Nemesis' planetary system with a solar outer planetary systems or of several encounters with inner planetary orbits or the Main Belt of asteroids or objects of the KBO belt? More gravitational modeling is needed keeping in mind a Nemesis dwarf star that crosses periodically through the Sun's planetary system. Better answers for the scattered disk objects (SDOs) may be available.

VII.

EFFECT OF NEMESIS'S
ELECTROMAGNETIC PROPERTIES

Increasing knowledge of electromagnetic phenomena and studies of interplanetary plasma are recently leading researchers to suspect that our Nemesis visitor is creating more kinds of havoc than originally thought. The effect of this brown dwarf's highly magnetic and electrical properties is now considered very prominent in affecting the inner solar system and possibly some of the outer planets as well.

Electrical arcing due to this magnetically and electrically charged star is produced and may be enhanced during any close passing or conjunction of one or more of the planets of either star. Planets shorten the electrical path length or plasma column and aid in increasing the flow due to their own electrical and magnetic fields. As the dwarf star makes its closest approach to the Sun both stars become electrically excited and exchange solar winds and plasma through streams of charged particles. If one or more planets are orbiting between these two bodies during the closest approach, an electric discharge of this plasma could very likely occur on the surface of an intervening planet causing great damage to its surface. Or, streamers of charged particles from the affected planet could trail the planet's orbit like a comet's tail. The geological effects of high energy discharge arcing are unlike any other such as crustal cracking or volcanism. Much electrical discharge sputtering of surface material is released into the atmosphere and then returns to the surface. Larger chunks of materials can gain enough energy to escape the gravity field of the planet and be carried along either in the orbit of the planet or with the conduit of plasma that is completing its quest for the next charged body of opposing electrical charge.

Another serious affect postulated by this article is the electromagnetic effect of Nemesis's strong magnetic field coupling with Earth's magnetic field which was possibly stronger than today. This coupling applies a torque to the Earth's already electrified crust and mantle which in turn increases the effect of plasma discharges arriving from Nemesis. The torque is transmitted to the mantle which creates an impact that breaks loose the magnetic connection between the Earth's lower mantle and the outer liquid core near the poles. The impact causes a brief shearing that rotates the mantle with respect to the much denser iron core. The process is labeled "pole shift" by current theorists. A partial pole shift occurred as is revealed by the difference between the current pole and magnetic axes (not currently popular) and by the dramatic changes in the northern and southern polar ice caps.

Any debris created by this brown dwarf passing between Mars and Jupiter is kept in disarray thus creating a permanent Main Belt of asteroids that can never be removed over millions of years due to gravity perturbations and solar winds as happens in other parts of the solar system. The supply of asteroids is continually disturbed and perhaps replenished every 3600 years not allowing the millions of years of time to clear and heal the planetary system of debris. In turn, this debris is then scattered into other parts of the solar system to create more havoc over longer periods of times long after the brown dwarf star has left the scene. This debris is either in the form of long period comets or asteroids in inclined and highly elliptical orbits. The question of where both short and long period comets come from is probably answered without the need for an unproven Oort cloud at 1½ to 2 light years away.

VIII.

THE STAR SYSTEM'S BARYCENTER

Another effect is the addition of Nemesis's gravitational field which changes the overall barycenter of the two-star system as the smaller star approaches its periapsis. The barycenter is the combined, changing center of gravity of all the bodies in the system that becomes the center of rotation for all the bodies. The Sun's increased wobble may induce abnormal tides and different heat output on Earth. In fact, the affect may over a short period of time while Nemesis is moving through its periapsis produce both unusually low and high tides and both unusually low and high temperature output of the Sun. These kind of short period changes are reminiscent of the recorded climate changes, droughts, and plagues that devastated Egypt during ancient times.

The Sun's barycenter changes due to an orbiting Nemesis is generally slow enough for the solar planets to adjust, except when the sister star approaches the periapsis and its orbital speed increases. Although the distance is much shorter between the stars there may not be enough time delay for the Sun's planets to maintain a constant distance from their star.

Let's briefly explore the possible changing barycenter of the Sun-Nemesis system. For a simple two-body problem the barycenter distance from the center of the most massive star, the Sun is "r". Then:

$$r = a \times [m_2 / (m_1 + m_2)]$$

where m_1 = the mass of the most massive body, the Sun; m_2 = the mass of the other orbiting body; and, a = the distance between the centers of the two bodies. Assuming from a previous determination Nemesis $\approx 0.052 \times m_1$ then the distances become $a_a \approx 473$ AU at its apoapsis and $a_{p1} \approx 40$ at its possibly farthest periapsis and $a_{p2} \approx 2.7$ AU at its closest periapsis. Then the calculated barycenters are respectively r = 24.6 AU, 2.0 AU, and 0.14 AU by letting $m_2 / (m_1 + m_2)$ approximate m_2 / m_1 since m_1 is much larger. In comparison to Jupiter which produces a maximum of r = 742,000 km which is only 1.07 times the Sun's overall diameter,[10] these postulated barycenter determinations are significantly larger. Due to Nemesis's mass and depending on the elliptical alignment of its orbit to the Sun's overall proper motion, these barycenter values reveal that the Sun weaves or wobbles dramatically especially as Nemesis gets closer to its periapsis. Due to feedback delay of the Sun's dense core motion and the Earth's orbit, making adjustments to compensate for the system barycenter being far outside the Sun's limb, solar radiation output and input to Earth can very possibly change abnormally. Hence, if no close encounters, collisions or electrical arcing occur in a 3600-year cycle of Nemesis' crossing, severe tidal acceleration forces can create minor crustal disturbances and ocean flooding; and, temperature fluctuations can cause climatic changes that can make regions either colder or warmer.

IX.

THE NEW NEMESIS HYPOTHESIS SUMMARIZED

Before delving into more details, the hypothesis of this paper will be briefly summarized. There exists a brown dwarf star orbiting our Sun whose orbit is extremely elliptical and travels most of its life outside the orbits of the Sun's planetary system. Periodically, or during each orbit, this star called Nemesis enters the Sun's planetary system and causes devastation of various degrees. Among the more important effects to Earth are changes in solar output from the Sun due to barycenter disturbances; increased tidal accelerations affecting earthquakes and volcanism; increased production of near-Earth asteroids of which some may collide causing global atmospheric dust and climatic changes; increased incoming plasma and magnetic field changes causing partial pole shift and attending geodesic adjustments with attending global flooding. The possible approximate orbital period is every 3600 years with the best-defined Nemesis visit and upheaval occurring 11,500 years ago. The effects of other recent visits are discussed later.

This sister star, or Nemesis as it is labeled by the scientific community, is due to come through the inner solar system within the near future and is touted as being observed currently by telescopes in the southern hemisphere. This information supplied by some "insiders in the astrophysics arena" is supposedly being suppressed by NASA and the government to avoid a global panic. But let's assume the data is correct and our Nemesis was already discovered and its orbit is plotted to confirm a gravitational connection to our Sun. It is predicted by numerous archaeologists that this star or planet sometimes referred also as Nibiru or Marduk in ancient texts returns every 3600 Earth-orbits or years. In fact, one claim is that some of mankind's mathematics is based on this number of orbits. That is why the ancient mathematicians measured time (the hour) and the circle by dividing them into 360 parts, multiples of the 3600-year orbit and the almost 360-day year. Let's go back in man's recorded history and global biological/geological records to verify anomalies or chaos that may have occurred at each 3600-year transition utilizing the best-defined boundary at 11,500 BP or 9500 BC as the starting point.

X.

STUDYING EARTH'S HISTORY AT THE BOUNDARY OF EACH INTERVAL OF EVERY 3600 YEARS

Going forwards and backwards from this marker or boundary of 9500 BC (11,500 BP) every 3600 years produces the following postulated boundaries of 4900 AD, 1300 AD, 2300 BC, 5900 BC, **9500 BC**, 13,100 BC, 16,700 BC, and 20,300 BC when the hypothesized Nemesis star passed through the solar system during its periapsis. Using the Jupiter period of 12 years as a basis the passing could take possibly from 6 to 30 years, but its effects could last for hundreds of years. The next future passing is predicted as **4900 AD,** which is comfortably far in the future. Nemesis has traveled in its orbit from its last periapsis for (2014 − 1300) = 714 years which means any emitted radiation is decreasing as it still recedes. At 3100 AD Nemesis will have reached its predicted apoapsis and will once again start to approach the Sun's planetary system.

A. THE LITTLE ICE AGE OF 1300 AD

The next interval going backward in time is **1300 AD**. The significant event of that time was the Little Ice Age that occurred from about 1350 to 1850.[11] Nemesis's passing during this time had no relatively large effect on Earth's climate or geological condition as occurred 11,500 years BP. Probably Nemesis either caused a wobble of the Sun reducing its heat output and/or several smaller asteroids were disturbed and sent inward having some strike Earth. Other official attributed causes besides an unexplained cyclic low in solar radiation are increased volcanic activity, changes in ocean circulation, and decreases in human population. This loss of heat output from the Sun is represented by the Maunder Minimum (1600 to 1750 AD) or lowest sunspot activity since modern times. Solar activity as measured by carbon14 dating indicated minimums during the 1300's (Wolf Minimum) and three other subsequent minimums.[12] The beginning of the Little Ice Age was marked by the growing Atlantic pack ice, killing of plants due to glaciation, stoppage of dependable summers in Northern Europe, and the Great Famine of 1315 to 1317.[13] The famine possibly weakened the health of the populations throughout Europe and Asia causing the Black Death or Plague of 1346 to 1353.

Any possible asteroid strikes were not lethal or recorded, but could have produced dust in the atmosphere along with numerous known huge volcanic eruptions to reduce solar radiation and cause a brief cyclic climatic cooling during this 500-year interval.

Surprisingly, no record is documented of observing Nemesis's passing. Of course, the telescope was not invented and used until the early 1600s. And, very cloudy conditions existed in the higher latitudes at that time. Regardless, this passing star should have made some visual impact. Perhaps some prior perturbation caused the perihelion to increase beyond the Jupiter and Saturn orbits making it difficult to be observed. This possible outcome bodes well for the future of the inner planets including Earth and its inhabitants. Nemesis at that time showed no serious direct effects on Earth; only indirectly did Nemesis possibly stir up some asteroids and cause the Sun to wobble from its normal barycenter.

B. INCREASED INTEREST IN ASTRONOMY OCCURS AT 2300 BC

Going backward in time again another 3600 years we arrive at **2300 BC**. During this period, an increased interest in astronomy was recorded in Babylon and Egypt. A religion based on observations of stars reaches its zenith in Egypt's Third Dynasty and was later replaced with Sun worship. From Chinese sources come the first recorded observations of a comet in 2296 BC.[14] This astronomical interest could have been stirred up by the arrival of new wanderers in the sky, namely Nemesis and one or more of its planets. Significant events did occur during 1600 to 1700 BC which could have been caused by Nemesis's previous visit of 600 years before. Once again asteroids and planetary satellites could have been disturbed and were sent inward toward Earth and other inner planets randomly causing impacts. Comet or asteroid strikes could have caused the following recorded changes in climate and resulting upheavals in man's civilizations. The Indus Valley Civilization ended in 1600 BC; the Shang Dynasty started in China about the same time; the last species of the mammoth became extinct in 1650 BC; in 1627 BC the cooling of world climate lasted several years as is recorded in tree-rings all over the world; major eruptions of the Minoan volcano, Thera, and Mount Vesuvius occurred during these same times. Egypt was conquered by Asian tribes; the Unetice culture located in the Czech Republic ended. The oldest astronomical documents were created by the early Babylonians during these times; a 21-year record of the so-called strange appearances of Venus was found in a Babylonian library that supposedly occurred close to 1600 BC.[15] Obviously, some outside factors were disturbing Earth's sky, climates, and its volcanoes. And, the stability of man's earliest civilizations was weakened. Man's interests in astronomical events came very much into focus probably due to the large, unexpected changes that occurred in an earlier remembered sky and by current terrifying comets streaking across the sky.

C. MAJOR FLOODING AND CONFIRMATION OF SAROS CYCLE IN 5900 BC

Subtracting another 3600 years from 2300 BC produces the year of **5900 BC**. Archaeologists in Scotland discovered evidence of a massive tidal wave that swept across the North Sea causing widespread destruction on the northern coasts of Europe citing 5800 BC as the date.[16] This dating is similar to the break-through of the Bosphorus Strait that caused major flooding of the Black Sea coastal areas. A massive volcanic landslide from Mt. Etna in Sicily caused a widespread tsunami on the coastline of Europe, Asia and Africa. The disappearance of Equids (species of horses) from the Americas occurred close to the same time. Occurring in 5677 BC a cataclysmic volcanic eruption of Mount Mazama created Oregon's Crater Lake.[17]

Besides 5900 BC indicating major geological events, flooding, and extinction events, man was beginning to record history in which this millennium was identified as one Saros or "Nibiru orbital" cycle occurring after the Great Deluge or the Noah Flood. Nibiru is a Sumerian "prophet-planet" that returns on a 3600-year cycle with the next cycle being 5900 BC.[16] The Sumerian/Babylonian word "sar" was one of the ancient units of measurement with a value of 3600.[18] The new Babylonian empire called Sumerian's Nibiru as "Adad" that was described in the Gilgamesh Epic as assaulting the Earth during the Flood. Some Jewish Old Testament experts also identify this time with the destruction of the Tower of Babel and the flight from the city Ur of the Chaldees and Abraham's family.[16] Apparently, the memory and history of the Great Flood was passed down 3600 hundred years earlier and was attributed to a God or star that the ancients called Nibiru/Adad. Of course, this paper now refers to these ancient names as Nemesis.

The story of the 5900 BC period is continued for several hundred years with the idea that Nemesis's visit invariably causes an aftermath of changes created by its tidal accelerations on Earth's crust and/or by its disturbance of some asteroids that strike Earth. Around 5600 the beginning of the desertification of North Africa caused migrations toward the Nile River valley starting the basis for the Egyptian civilization. Irrigation and the beginning of the Sumerian culture started in earnest about 5400 BC.[17] Other major world-wide Neolithic cultures developed in less than two centuries.

After certain climatic disturbances began to lessen and cataclysmic geological events ended, mankind was then ready to leap forward to create major cultures world-wide near the end of the Neolithic Period and during the Chalcolithic Period ranging from 4000 to 3000 BC. It should be noted that around 3200 BC various indicators revealed global changes in climate supposedly caused by a drop in solar energy output. Some of these indicators were a quick shift in the formation of ice caps in the Peruvian Andes; tree rings in the British Isles showing its driest period; unusual changes in plant pollen uncovered from lakebed cores; lowest levels of methane retrieved from ice cores in both Greenland and Antarctica.[19] This paper is not attributing this global climatic change to any Nemesis visit directly. Perhaps some random immense volcanic activity and/or asteroid impacts caused long term atmospheric dust to cool Earth in certain regions.

D. THE GREAT FLOOD EVENT OF 9500 BC

Continuing in the same manner by going backward another 3600 years or one orbital period of Nemesis or one Mesopotamian "sar" from 5900 BC produces the infamous age of **9500 BC** when a certain cosmic catastrophe and the "Great Flood" occurred. This period also is the ending of a significant geological and climatic period called the Younger Dryas. If indeed man had developed advanced civilizations prior to this time such as is proposed by Plato's Atlantis and the land of Mu then it was thoroughly destroyed with only memories left to be verbalized.

Directly following this time starts the Mesolithic Period lasting about 1000 years. Actually, the Mesolithic Period is dated by the type of hunter-gatherer tools used and varies throughout parts of Eurasia.[20] The last Stone Age stage is the pre-pottery/pottery Neolithic Periods that saw small farming communities beginning to form with a slow progression of domesticated crops and animals. The rise of this Neolithic era is seen in Southeast Asia, the Fertile Crescent, Europe, China, and Korea[21] – all in the middle northern latitudes where most of the megalithic structures are located. Were there developed civilizations with megalith structures before the Flood? If yes, then the survivors had to rebuild infrastructure and reconstruct reasons for the megaliths with much of their knowledge lost?

The beginnings of a developed culture matching the later Neolithic stage in the eastern Mediterranean (Levant region) around Jericho and Palestine are dated closely to 8800 BC.[21] This time is unusually early compared to other Neolithic cultures. Perhaps destruction in the Levant region was mild and the already developing Neolithic culture was saved. For this reason, the "Cradle of Civilization" soon arose giving us the written word about the Flood. People of the Americas and Pacific regions mostly retained the Neolithic level of tool technology until the time of European contact. Climatic changes of the previous Younger Dryas Period are thought to have forced people to develop farming.[22] Or, was farming already developed prior to the Flood? Perhaps the farming and urban culture with polished stone and metal tooling already existed in certain areas of the world and was driven backward toward Mesolithic adaptations.

E. THE OLDEST DRYAS PERIOD MARKING THE NEXT TWO SAROS CYCLES

Going back farther in time leads us into the colder parts of our current Ice Age. These parts are the stadial periods of lower temperatures during interglacial warm periods that separated the glacial periods of the various ice ages. The immediate three colder stadial periods are named the Oldest Dryas, the Older Dryas, and Younger Dryas Periods. The approximate endings of two of these periods agree with the two orbital periods of passing Nemesis: **9500 BC** at the end of the Younger Dryas Period; and 9500 BC minus 3600 years = **13,100 years BC (15,100 BP)** at the end of the Oldest Dryas/Older Dryas Period. The last Nemesis visit to contemplate is 13,100 BC minus 3600 years = **16,700 years BC (18,700 BP)** which is the beginning of the Oldest Dryas.

These Dryas periods are named after an indicator genus, an arctic and alpine flowering plant Dryas. Their fossils are found in higher concentrations in geological deposits between these colder stadial periods. The Oldest Dryas Period began between 19,000 and 18,000 years BP and ends with a fairly high resolution at 14,600 BP.[23] The beginning and end of the Oldest Dryas Period are well identified by post-glacial sea level rise. The meltwater pulse marking the end of the Older Dryas lasted about 1000 years with sea level rising over 20 meters.[24] The middle or Older Dryas Period is difficult to predict with a good resolution of its beginning time which is probably due to the Earth still recovering from the previous Oldest Dryas stadial period. The various dating methods for the Older Dryas period provide a range lasting only 100 to 150 years and centered near 14,100 BP.[25] Two smaller well defined interstadial periods or oscillations occurred one before and one after the Older Dryas Period. These are called the Bolling and Allerod oscillations respectively. The Older Dryas and the two oscillations span 13,000 to 11,000 BC.[26,27]

A discussion of general astronomical cycles that occur on Earth is appropriate at this juncture in this paper. Cycles of a year or less including the 11-year sunspot cycle are not considered. For the most part all these larger cycles have much longer time intervals than the 3600-year cycle that is being postulated for Nemesis. This comparatively short cycle matches the outer planets and Kuiper Belt objects orbiting the Sun such as Pluto with an orbital period of 249 years and Sedna with an estimated period of 11,400 years. All other known and studied cycles are much larger such as the Earth's precession at 26,000 years; the obliquity or change of Earth's tilt every 41,000 years; the eccentricity of the Earth's orbit every 100,000 years. These Milankovitch cycles help to determine the long periods of glaciation on Earth.[28,29,30,31] And, then there are the extremely long periods of 26-million-year mass extinctions. Obviously, these cycles cannot be correlated with Nemesis. However, one recent series of short cycles was discovered by analyzing the ice cores of Greenland. Rapid shifts of about every 3600 years from peak to peak in isotopic composition occurred from a span of 10,000 to 100,000 year ago. This isotopic study called the Dansgaard-Oeschger Events measured the frequency of stadials and interstadials which are relatively cold and warm periods, respectively, occurring between much colder and longer glacial periods. This Dansgaard study did not attribute the cause, but this paper suspects Nemesis's influence.[32]

It must clearly be understood that some events caused by the passing of Nemesis may lag by several hundred years due to new asteroids being created and/or disturbed from stable orbits that eventually strike Earth as large bolides. If the bolides are large enough, resulting global dust placed into the atmosphere can cause unusual cooling for decades to come. Also, tidal surges of Earth's crust due to a close passing of Nemesis and/or its conjunction or opposition with the Sun may create cavities for lava flow within the lithosphere resulting in a delayed volcanic activity and increased movements of plate tectonics.

Table 14: The Postulated Cycles for Nemesis

AD or BC (years)	BP (years)	Boundary of Cycle or Return of Nemesis (every 3600-year interval)
4900 AD	≈ 2900 years into the future	Nemesis's next predicted return is several centuries in the future. Currently, Nemesis is receding from the Sun.
1300 AD	≥700	Nemesis's last return triggered the Little Ice Age and possibly increased volcanic activity reducing solar heating.
2300 BC	4300	Interest in astronomy increased and climate changes were triggered 700 years later about 1600 BC.
5900 BC	7900	Massive flooding and extinction events marked this boundary; the Sumerians marked this boundary as the next Saros cycle from the Great Flood event.

AD or BC (years)	BP (years)	Boundary of Cycle or Return of Nemesis (every 3600-year interval)
9500 BC	11,500	During this time, the Great Flood and other catastrophes including mass extinctions occurred; this boundary is also the beginning of receding glaciation and the end of the Younger Dryas geological period. A sea level rise started that attained present sea level about 5000 BC. The brief rise during the Great Deluge that probably spanned several hundred years is questionably not discovered.
13,100 BC	15,100	This boundary marks the end of the Oldest and Older Dryas stadial periods with two interstadial periods occurring close together. A meltwater sea level pulse started and lasted about 1000 years.
16,700 BC	18,700	This boundary marks the beginning of the Oldest Dryas period when the Earth was departing from its last glacial maximum.

XI.

IMPROVEMENT OF TIME BOUNDARY ANALYSIS FOR RETURNING NEMESIS

In May of 2017, I was honored and privileged to receive a brief, but very important peer review of my Nemesis hypothesis. A person connected with various NASA programs and trained in astrophysics provided concise and sincere advice about a main point made in this paper. My plotting of the data of historical events, both mankind and natural, of the postulated cyclic events of the returning Nemesis star to the inner solar system every 3600 years was criticized. The paper uses as the main datum point, the end of the well-defined Younger Dryas geological period which is 11,500 years BP. The paper then continues to subtract or add 3600 years from that point in time to determine other visitations of the Nemesis star and look for disturbances on Earth at those times. The major fault revealed is that no stated research used a continuous plot of major Earth events between those points in time. Per the criticism, it is an easy matter to look for any events occurring during certain peak times and ignore all others; the logic is flawed and no real conclusive evidence is actually gained. Yes, I agreed that a more continuous plot of data is required.

The predicted times for Nemesis's visits to the inner solar system are the years before present (*BP*): 18,700; 15,100; 11,500; 7900; 4300; and 700 (1300 AD). The next arrival is predicted for 4900 AD. So, any continuous kind of record going back 20,000 years or even 11,500 years ago becomes scantier the farther back one goes. Checking the history of volcanoes from www.randomhistory.com reveals a random occurrence going back to 1600 BC. Going farther back reveals more random volcanoes to about 4000 years by using the *Timeline of Human Prehistory*. However, during the 8th millennium (7911 BC) seven massive volcanic eruptions occurred lowering the global temperature for several centuries as is revealed by Greenland ice cores. Other major known, dated volcanoes occurred in the 7th millennium (6600 to 6100 BC): the 900-km² lava fields of Iceland, the Kurile volcano on Siberia's Kanchatka Peninsula; and volcanic fields in central Washington State. The 6th millennium (5677 BC) had eruptions of Mount Etna in Sicily and Mount Mazama in Oregon creating Crater Lake, the largest eruption in the Cascades. These eruptions more than likely resulted from the continued settling of Earth's geoid displacement during the Great Deluge Event of 11,500 years BP. However, the next predicted visit of Nemesis in 5900 BC could have caused some further gravitational and magnetic adjustments of the Earth's crust at that time. In fact, one difficulty of dating ice cores is the lack of known volcanism to be used as markers that occurred any earlier than 8500 years BP.

The source, www.randomhistory.com, also lists major earthquakes and tsunamis. The oldest, super tsunamis are predicted to result from an asteroid that struck the Indian Ocean causing waves 600 ft. high; and the sudden tectonic plate movement that caused the tsunami in Crete and surrounding Mediterranean coasts in 1530 BC. Other tsunamis from the *Timeline of Human Prehistory* indicated a super tsunami, the Storegga slide in the Norwegian Sea in 6100 BC and another super tsunami in the Eastern Mediterranean thought to be caused by Mt. Etna's eruption. This search for earthquakes and tsunamis gives a similar result of random occurrences and nothing documented or discovered earlier than about 4000 years except for the previously mentioned super tsunamis. The randomness of earthquakes and tsunamis is probably the result of small tectonic plate movements since 80% occur in the Pacific Ring of Fire. One of the reported major earthquakes occurring in the eastern Mediterranean around 1201 AD corresponds closely to 1300 AD predicted visit of Nemesis. Major volcanoes occurring near this date are Mt. Vesuvius in Italy at 79 AD and Hatetepe in New Zealand at 180 AD. But there are no conclusions to be made about Nemesis's visit in 1300 AD or any other visit during mankind's prehistory by studying the data of volcanoes, earthquakes, and tsunamis. Also, no conclusions can be made about the history of hurricanes which only goes backward to 1900 as revealed by www.nhc.noaa.gov/outreach/history. So, what other historical data can be analyzed to search for some periodic cycle for the orbiting Nemesis?

By studying the rise and fall of interlocking civilizations and the beginnings of major cultural developments such as pottery, proto-writing, the Copper, Bronze and Iron Ages may reveal when major disturbances on Earth caused the downfall or required rise of replacement cultures and different techniques for surviving. Again, the *Timeline of Human Prehistory* and also the *Timeline of Ancient History* are consulted. Some the earliest dated human developments are listed:

- 200,000-180,000 years ago: Time of mitochondrial Eve & Y-chromosome Adam.

- 195,000 years ago: Oldest homo sapiens fossils.

- 70, 000 years ago: cave wall abstract art and personal adornments.

- 64,000 years ago: bow and arrow following the spear.

- 50,000 years ago: sewing needle

- 42,000 years ago: Paleolithic flute and high-level maritime skills in East Timor
- 40,000 years ago: figurines
- 29,000 years ago: ovens
- 28,000 years ago: twisted rope
- 25,000 years ago: huts built of rock and mammoth bones
- 20,000 years ago: harpoons, saws, oldest pottery
- 15,000 years ago: domestication of the pig
- 12,000 years ago: domestication of sheep and goats
- 11,000 years ago: construction of ceremonial sites
- 10,500 years ago: domestication of cattle
- 10,000 to 9000 years ago: barley and wheat along with bread and beer begin
- 7500 years ago: invention of wheel and proto-writing and copper smelting
- 6000 years ago: domestication of horse and chicken with civilizations developing around the Fertile Crescent of the Middle East
- 5300 years ago: Bronze Age begins
- 5200 years ago: writing is invented

What is revealed in this brief, abbreviated list is that humans were well developed prior to antediluvian times. The Great Deluge event with all its utter destruction caused man to start almost all over again around 11,500 BP. However, as can be easily seen, his skills for both survival and a vibrant culture were already honed.

The radio carbon dating for these discoveries in anthropology and archeology are questionable due to natural historians having a strict paradigm of human culture progressing forward from the stated Stone Age period called the Neolithic Revolution that supposedly occurred around 11,000 to 9000 years BP. Of course, mankind's technological evolution is assumed to have started at that time because previous technologies and their required knowledge were obliterated during the Great Deluge. If so-called anomalies in the dating method occurred that led much farther back to 20,000 or 100,000 more years, historians would dismiss or even banish this information as being just errors in laboratory techniques of dating. No one can really know for certain at what stage man evolved prior to antediluvian times or prior to 11,000 years BP. Keep in mind, that many of these digs where artifacts of tools and art were found, are very distant and isolated from any ancient urbanization where education could be directly passed on. The implication is that advanced knowledge was well implanted worldwide for the majority of humans wherever they existed.

The challenge is still to find cycles for Nemesis's crossings into the solar system. The *Timeline of Human Prehistory* that shows man's progress from 11,000 years BP to about 5,500 BP covers the time from the Middle Paleolithic (Old Stone Age) to the beginnings of the Bronze Age. No periodicity of about 3600 years can be found in the listing of the ages of mankind's settlements, cultural developments, and artifacts. However, the constant rate of development does not really exist. After the Great Deluge, there is little chance for any possible easy communications or transportation of people; languages and cultures become disconnected, and progress at different rates with many distinctions occurs. Suspicions and paranoia are created when any of the isolated cultures meet up thereby delaying progress in many regions.

The *Timeline of Ancient World History* is a collection of historical events that goes from the 10th millennium BC (12,000 years ago) to the 4th century AD. Starting with the 40th century BC the data becomes more populated for each century as time moves forward. This documented timeline goes from the beginning of recorded history to the Early Middle Ages and includes the Bronze and Iron Ages. The source eh-resources.org/timeline-middle-ages/ extends the analysis of timelines into the Middle Ages and the Early Modern Period. What is essentially revealed is a random continuum of beginnings and endings of civilizations, large migrations of people, droughts, animal and plant domestications, and technological/cultural developments. However, when one connects the tracking of changing climatic conditions with the various collapses of civilizations, cyclic periods do appear. These periods of climatic maladies actually pop out to grab your attention. Many recent studies have been made in these areas of historical climatic conditions because of the present concern for global warming and unwanted, accelerated rise of sea level. These studies consist of the "kiloyear events" that were assembled with the combination of global temperature changes from Gisp2 Ice Core Data of Central Greenland and other ice core data of mountain regions, sedimentary records, aridification records, methane concentration in the atmosphere, hemispheric cold snaps, sea level changes, and the collapse of various civilizations caused mainly by drought.

Another set of climatic fluctuations in the Holocene Epoch called the Bond events attempts to identify a 1500-year cycle. These events are mainly based on petrologic tracers of drift ice in the North Atlantic. Gerard C. Bond of Lamont-Doherty Earth Observatory at Columbia University tried in vain to establish such a cycle, but lacks a clear climate signal where only certain peaks correspond with periods of cooling and other peaks are only coincident with aridification of large regions. Mr. Bond failed to attach his cycle to some kind of solar cycle, and lacked an adequate model for encompassing all the known climatic, atmospheric, and geological fluctuations. Bond's postulated theory is currently claimed to be a statistical artifact with no cause or effect and has been rejected. The only Holocene Bond event that has a clear temperature signal in the Greenland ice cores is the 8.2 kiloyear event.

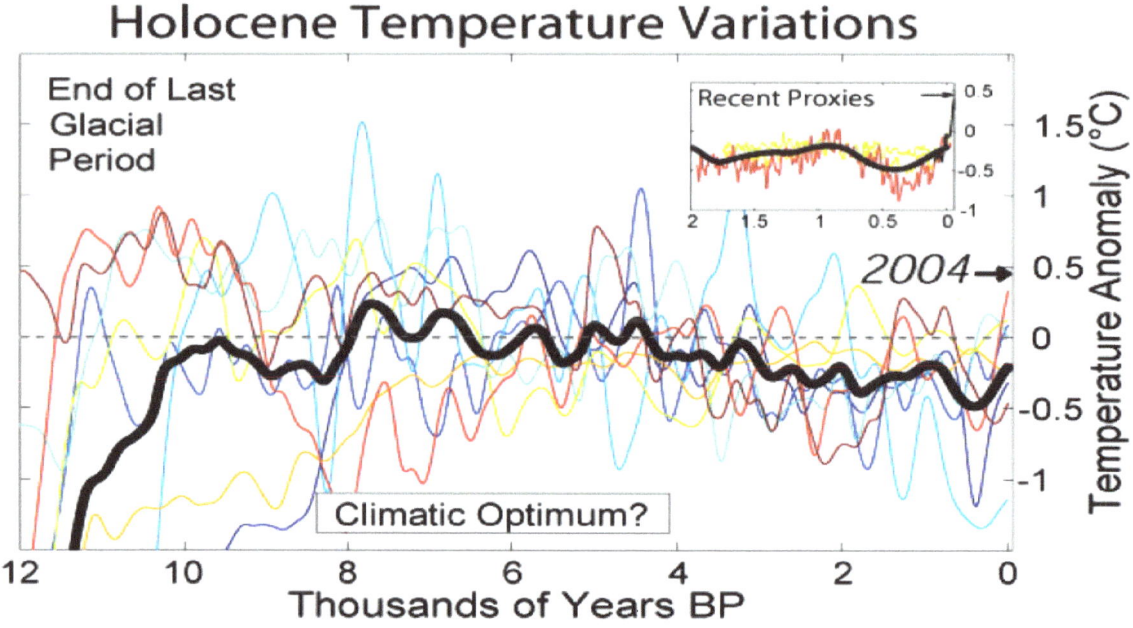

Figure 15: Holocene Temperature Variation
(Gerard Bond unsuccessfully postulated a 1500-year cycle using ice core
data going back 12,000 years ago.)

This paper is also trying to identify a certain periodicity, but this time the cycle is an average of 3600-year span of time due to the orbiting Nemesis brown dwarf crossing through the solar system. However, the periodicity is more complicated because this model offers not only a mean period, but also variability in its time range of affects due to electrical/magnetic phenomena created by the two-star system. This model is relying on the always changing electrical charge of the two stars. As the brown dwarf with its own magneto-sheath crosses the magneto-sheath of the Sun at least twice during one orbit, an exchange of electrical charge takes place in different amounts thus affecting the outpouring of solar wind in order for the Sun's system to reach equilibrium. The solar wind fluctuations in turn affect the anodic or negatively charged planets of the Sun including Earth. Also, due to the Sun possibly increasing or decreasing its electrical charge the orbits of the planets have to make minor radial adjustments which have little to do with gravitational forces.

The fluctuating solar wind and minor changes in orbital radius of the Earth both affect the weather and climate of Earth. Also, changing magnetic and gravitational influences will affect the geological stability of the Earth's crust in different and random ways.

If the orbit of Nemesis is highly inclined to the ecliptic plane of the Sun's planets it not only crosses the Sun's magneto-sheath twice in a postulated 400 ± 50 - year cycle to possibly exchange electrical charge, but also crosses the ecliptic plane twice in the region of the Main Belt of asteroids situated about 2 to 3 AU from the Sun. This intersection may cause some major perturbations of various minor-sized celestial bodies. So, the hypothesis of this paper not only includes a 3600-year cycle, but a 400 ± 50 - year cycle centered on the peak of the larger cycle. The effects of each cycle are different due to varying exchanges of charge between the two stars; the varying exchanges of charge between the two stars and their planets causes fluctuating solar winds. Slight adjustments of planetary orbital radii cause different inputs of radiant energy that either increase or decrease.

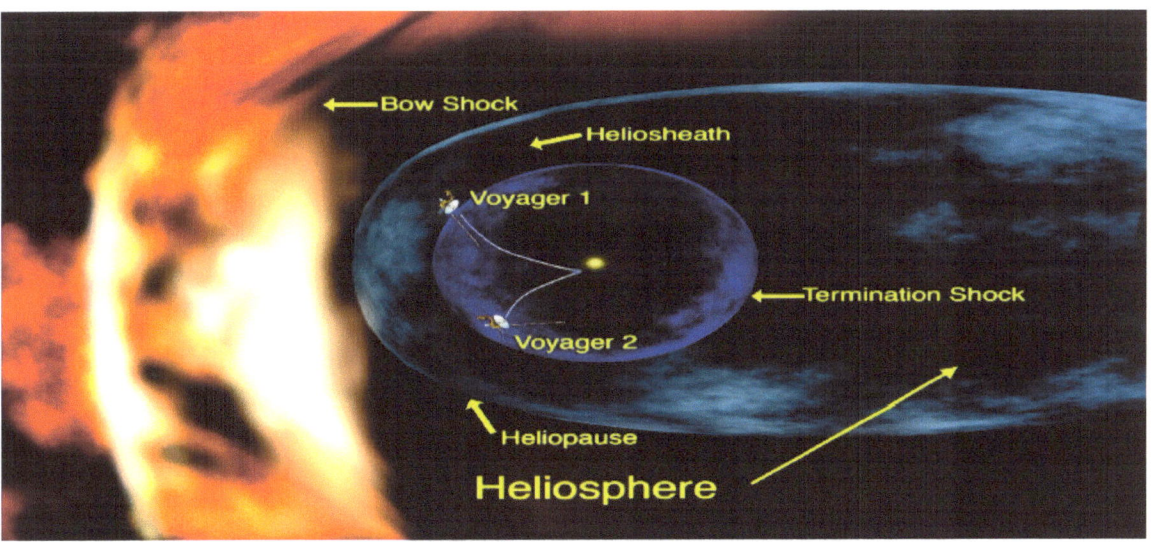

Figure 16: Heliosphere and Helio-sheath Boundaries Identified by the Voyager Space Probe Missions

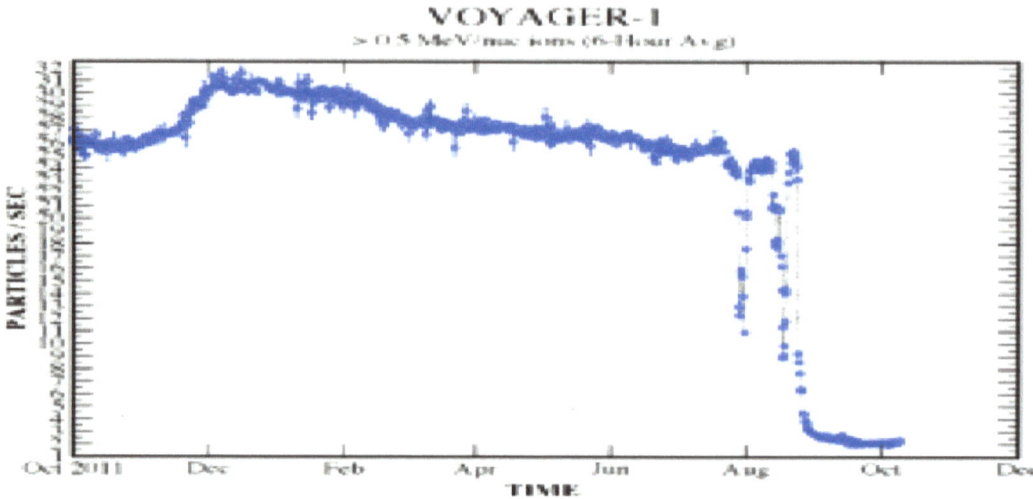

Figure 17: Voyager-1 Space Probe Data as It Approached the Helio-sheath

Figure 17 shows how the solar wind particles are captured by the Sun's helio-sheath that in turn provides a double-layer current for collecting external interstellar particles and the means for re-supplying the Sun's energy at its poles. If the Nemesis star crosses this sheath of current, then power can be transferred in either direction depending on the requirements of system's charge equilibrium.

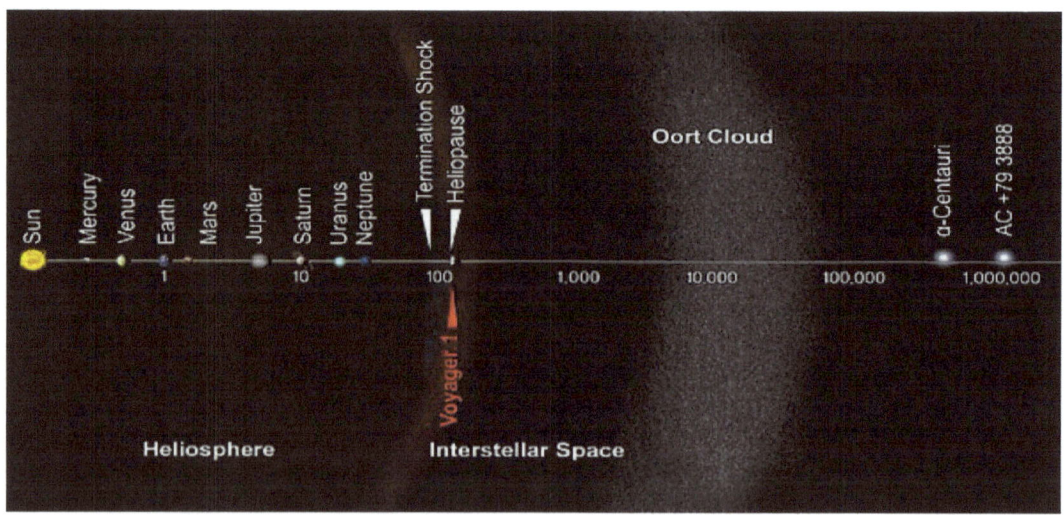

Figure 18: Location of the Heliopause or Helio-sheath at 100 AU from the Sun

(This distance is over twice as far as the farthest planet, Neptune.)

Henceforth, the Earth endures these cyclic changes in the solar system through changing weather, climate, and minor crustal adjustments. No pure cycle can be identified with any certain cause. The only chance of reckoning is to identify a possible mean cycle of 3600 $\pm$ 100 years with a full range of affects that occur at each peak of this cycle for a span of 400 $\pm$ 50 years. The accuracy of these predictions is based on a reasonable exchange of charge between the two stars and between these stars and their planets, the perturbations of the orbits, the precession of orbits, and orbital radii adjustments. These predictions cannot be solely based on gravitational considerations which consensus scientific thinking does. The more powerful, still mysterious, electrical-magnetic forces between celestial bodies is what maintains stability. Now, let us look at how all these changes on an interstellar and planetary scale affect Earth by examining the following listed kiloyear events.

The best-defined climatic fluctuation is the 8.2 kiloyear (8200 years ago) event that followed the Great Deluge event of 11,500 years BP. There was a sudden dramatic cooling and drying on a hemispheric scale as shown by the Central Greenland ice core data. The temperature drop was not as severe as the Younger Dryas cold period, but more severe than the Little Ice Age. The duration was about 150 years. There was an emission reduction of 15% in atmospheric methane, and CO_2 was lower by about 25 ppm over 300 years. Drier conditions started in North Africa and persisted for a 300-year aridification and cooling period. As the atmosphere is cooled dryer conditions prevail because the water vapor is quickly condensed retarding the Earth's water cycle of ceaseless evaporation and condensation. These drier conditions provided a natural force for Mesopotamian irrigation-type agriculture and surplus production to achieve the classes of people found in ancient urban life. These pressures on human agrarian culture led to more evaporation, drying, and soil erosion. The 8.2 kiloyear cooling is attributed to the meltwater pulse that became permanent. The sudden rise in sea level is ironically, but supposedly caused by the melting and collapse of the Laurentide Ice Sheet and drainage of Lake Agassiz-Ojibway in Canada. The irony is how a cooling period could collapse the existing ice sheets and raise sea level; data from the Rhine-Meuse Delta indicates a rise of 6 to 13 feet. Similar sea level data shows similar rises in the Mississippi Delta, northwest Europe, and Asia. This sea level data may be confused with oceans still receding after the Great Deluge. The remaining ice sheets that were moved southward after crustal/mantle displacement, continued to melt due to their new warmer latitudes regardless of the overall global cooling. The story of the Laurentide Ice Sheet moving southward during the Great Deluge is a missing link for consensus science.

For this author, it is very suspicious that in 2003, the Office of Net Assessment (ONA) at the United States Department of Defense was commissioned to produce a study on the likely and potential effects of a modern climate change. The study under ONA head, Andrew Marshall, modeled a possible climate change based on the 8.2 kiloyear event. Do "people in the know" want to keep as classified data of any information being gathered about the Great Deluge and its lengthy aftermath? Definitely, the academic community was not trusted in performing this 'net' assessment of mankind's real genesis and revelation.

The conclusion for this paper is that the 8.2 kiloyear event meets the conditions for the next period of return for Nemesis predicted at 7900 BP. The effects of another star such as Nemesis crossing the helio-sheath of the Sun every 3600 ± 100 years seems very plausible. On its entry into the solar system, another span of 400 ± 50 years later is postulated for Nemesis' crossing the helio-sheath once again before leaving the solar system. As the magneto-sheaths of the two stars cross each other, massive electron exchange can go in either direction for charge parity to be achieved. If the Sun loses electrons to the brown dwarf than the solar winds toward Earth are reduced; if the Sun gains electrons then the solar winds increase toward Earth. Other more shorter-term effects are produced by the perturbations of various planetary orbits and the barycenter adjustments between the two stars. These shorter-term effects vary widely in severity depending the orbital locations of the various planets with respect to each other and the two stars. Hence, weather/climatic fluctuations, glaciation/cooling periods, and even geological disturbances on Earth will be affected, but not in any perfect periodicity or amplitude. The only real periodicity is the mean average of the 3600-year orbital period that uses the Great Deluge event of 11,000 years BP as one sharp and well-established datum point.

The 5.9 kiloyear event is the beginning of intense aridification events of the Holocene which started the desertification of the Sahara and much of Western Asia. Unlike the 8.2 kiloyear event, it did not have temperature markers in the ice core data or other markers such as methane and CO_2 fluctuations in the atmosphere. Unlike other major kiloyear events it was not followed by any significant recovery. Its effect has continued relentlessly to present times. In the preceding millennia before 3900 BC or 5900 years BP Neolithic humans introduced domesticated animals and agrarian culture that may have played a significant role in stripping vegetation that caused cascading effects of both drier weather and climate. This labeled 5.9 kiloyear event is **not** considered as marking the return of Nemesis; this event is merely the onset of serious aridification due to continued cooling and drying by natural weather forcing and humans destroying the environment on the Arabian Peninsula and in eastern Africa. The subsequent drier atmosphere led to further degradation farther westward into northern Africa.

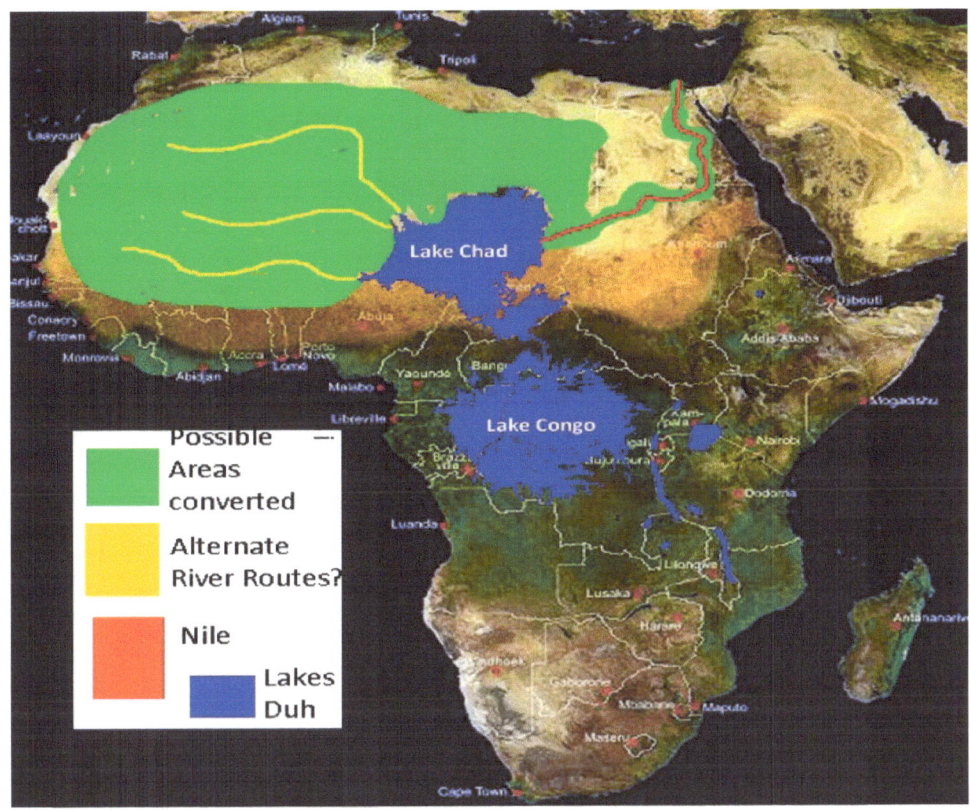

Figure 19: The Predicted "Wet Africa" before Major Aridification

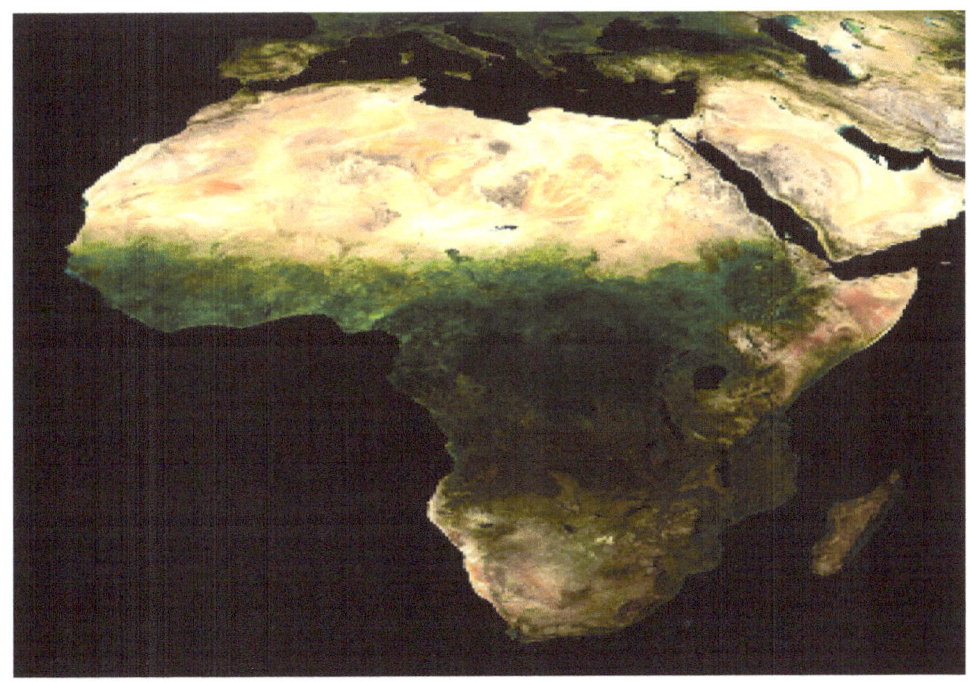

Figure 20: Satellite View of Africa Showing Present Aridification

The above figures show the difference of a 'Wet Africa' or the Neolithic sub-pluvial period of wet and rainy conditions in the Sahara about 7500 to 3000 years BC verses the dry Northern Africa and Eastern Asia of today shown by satellite photography.

The 4.2 kiloyear event was the most severe aridification of the Holocene period starting about 2200 BC. The event is hypothesized to cause the collapse of the Old Kingdom in Egypt, the Akkadian Empire in Mesopotamia, and the Liangzhu culture in China due to serious droughts. Also, archeological data reveals a significant southeastward migration of the Indus Valley Civilization. The claim is that no adequate signal in the ice core date of Central Greenland supports the 4.2 kiloyear, although the graph shows prominent drops in temperature between 4800 years and 4000 years BP with an average rise between these times. The expected return of Nemesis at 2300 BC is supported by this 4.2 kiloyear event. The temperature drops measured in the ice core data should be expected since the amount of solar wind and radiant heat energy could have been affected by Nemesis crossing the Sun's helio-sheath at the predicted 200 years before and 200 years after the mean orbital return of 4300 years BP. The lowest temperatures in the ice core data were measured at 4800 to 4000 years BP, a span of 800 years instead of the predicted 400 years. Of course, the actual recovery of Earth's climate will take longer than the predicted span of years simply based on the different orbital locations of Nemesis.

Another major event is attributed partially to climatic fluctuations. Historians call this event the Late Bronze Age collapse occurring between 1200 and 1150 BC. The Late Bronze Age broke down into isolated village cultures throughout the Near East, Aegean Region, North Africa, Caucasus, Balkans, and the Eastern Mediterranean. The possible causes are both environmental and cultural, but hypothesized to be mainly caused by a general system collapse of mankind's civilizations. The growing complexity and specialization of political, economic, and social organization in Carol Thomas and Craig Conant's words together made the organization of civilization too intricate to reestablish piecewise once disrupted. Certain flaws such as top-heavy political structure, revolt of the peasantry, and defection of mercenaries, crop failures, drought, and interruption of maritime trade caused the inevitable destruction of major cities across the land. This collapse was primarily caused by man's ineptness and has no connection to a visitation of the Nemesis star.

The next and last visit of Nemesis is predicted to occur about 1300 to 1400 AD. This period of environmental upheaval and climate fluctuation is called the Little Ice Age. The temperatures per ice core data dropped on average about 1.5 degrees Celsius which was corroborated by tree ring data. The cooling trend moved from north and west to south and east through Europe toward the Mediterranean. This cooling led to crop failures and famine from 1314 to 1317 AD. Eventually these pressures on society created the Black Death in

1347 that pushed the decline in population by as much as 40%. The Little Ice Age started about 1300 AD with its most severity from 1600 to 1800 and ended about 1870 AD.

The ice core data from Central Greenland is used as a direct comparison for the other kiloyear events. Although not called a kiloyear event, the Little Ice Age had all the same characteristics. The ice core data showed the temperature dropping from 1100 AD to its lowest point in 1300 AD which then quickly raised to normal levels before plunging again to low levels between 1450 and 1700 AD. The data reveals support for the hypothesis that the visiting Nemesis crossed the Sun's helio-sheath around 1100 AD and caused the solar wind to fluctuate creating cooler, drier weather on Earth. Then Earth's weather began to improve slightly until the orbiting Nemesis star travels from its periapsis and once again passes through the helio-sheath to affect change in solar wind conditions. When the Nemesis star leaves the solar system boundary and heads for its apoapsis, it may still take 100 or more years for Earth's recovery.

Further proof of the Sun being affected electrically during this last visit is the Maunder Minimum which coincided with the coldest part of the Little Ice Age. Before this time there were few records about sunspots. E.W. Maunder discovered the absence of sunspots which now is known to mean a less active and colder Sun with less energy output to heat the Earth. Also, some recent published data supports the idea that the Sun expanded and slowed its rotation. In the Electric Universe scheme of things, this means that electric current was drained off by the Nemesis star as it crossed the helio-sheath which collects electrical energy from galactic space and supplies the Sun at its polar regions. This supply of electrical energy via the helio-sheath was disrupted and the Sun's energy input was reduced. In order to maintain electrical-charge equilibrium the Sun ejects in varying amounts of energy in the form of solar wind which in turn supplies energy to the Sun's planets. For Earth this energy is both in the form of radiant heat energy and electrical energy which drives weather and climate. Both the sunspot minimum and the global cooling of Earth are proof that the Sun's energy supply was reduced and postulated to be caused by the crossing of Nemesis through the solar system.

In conclusion, the studies of kiloyear events and their effect on Earth's climate prove substantially the cyclic nature of a possible Nemesis brown dwarf orbiting its sister star, the Sun, every 3600 years. But where is Nemesis? Perhaps it was seen in the 1300's but was confused with one or more comets. The use of telescopes and more serious study of the sky were to come later. The mystery as to why a sister star cannot be seen now, will be addressed next. The analysis is summarized in the following table listing the correlation of the Nemesis' 3600-year Sar cycles and kiloyear events. The Little Ice Age and the Great Deluge events are considered kiloyear events. The so-called 5.9 kiloyear event and the Late Bronze Age collapse of 1200 BC do not qualify as true kiloyear events as previously explained.

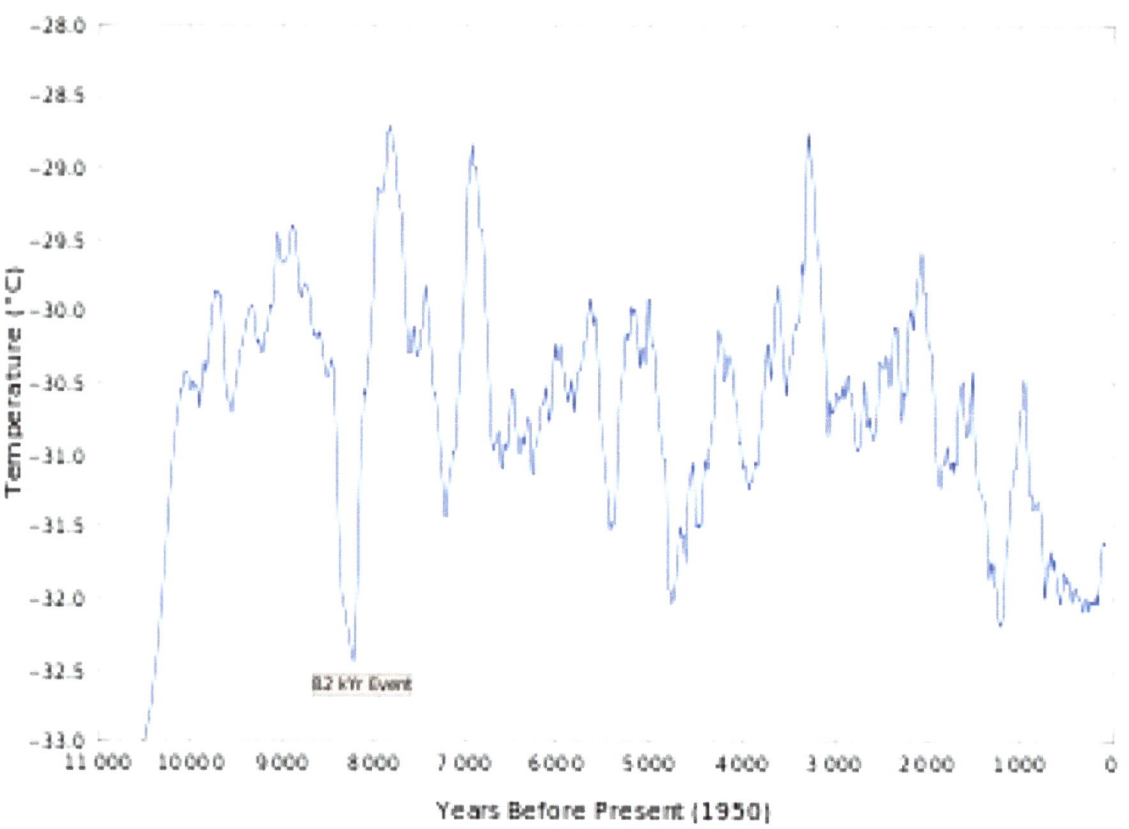

Figure 21: Central Greenland Ice Core Data (Gisp2) Reconstructed
Temperature
(The Gisp2 data shows unusual temperature drops for the 8.2 kiloyear,
the 4.2 kiloyear, and the Little Ice Age events.)

Table 22: Summarizing the Comparison of the Predicted Nemesis 3600-year Sar Cycles with the Kiloyear Events

Predicted 3600-Year Sar Cycles of Orbiting Nemesis Star (BP / BC)	Kiloyear Events for Major Climatic Changes Per Academic Studies (BP / BC)	Years of Lowest Central Greenland Ice Core (Gisp2) Temperatures	Full Range of Years for Actual Effects on Earth (Predicted max. is 550 years*)
11,500 / 9500	11,500 BP / 9500 BC Great Deluge or End of Younger Dryas	11,500-10,500 BP	Used as baseline.
7900 / 5900	8200 BP / 6200 BC (called the 8.2 kiloyear)	8200 and 7200 BP (with raised levels between these years)	7900 vs. 8200 BP; actual large fluctuations of 1200 years** are indicated.
4300 / 2300	4200 BP / 2200 BC (called the 4.2 kiloyear)	4800 and 3800 BP (with raised levels between these years)	4300 vs. 4200 BP; actual 1000 years** of large fluctuations.
700 / 1300 AD	800 BP / 1200 to 1800 AD (called the Little Ice Age)	800 AD and 1300 to 1800 AD (with raised levels between 800 and 1300)	1300 vs. 1200 AD; actual 500 years** of large fluctuation and continuing for another 500 years.
+ 2900 / 4900 AD	N/A	N/A	Strong effects could occur as early as 4350 AD.

*The predicted full span of years of disturbance is obtained by adding the 100 years of uncertainty for the Sar cycle of 3600 ± 100 years to the span of visitation by Nemesis of 400 ± 50 years or 100 years + 450 years = 550 years.

**The actual years of disturbance is estimated by using the ice core data from Gisp2 shown in Figure 21. This method of only using one parameter of climatic change leads to inaccuracies, but aids in seeing how spans of time of Nemesis' visitation close to and inside the Sun's helio-sheath need to be considered.

Note: No lag for recovery of Earth's climate is considered; factors for variance in lag may be Sun's output of energy; orbital perturbations; geological disturbances such as volcanism with subsequent altered atmosphere, and glaciation fluctuation with subsequent altered sea level. All these factors can create sizable shifts in the climatic indicators. But what stands out in this study is the close resemblance of a 3600-year Sar cycle.

XII.

AFTERMATH OF THE YOUNGER DRYAS PERIOD

The Younger Dryas Period occurring between 12,800 and 11,500 BP[33] is the best documented and most important stadia period for consideration by this paper. The 3600-year marker for one of Nemesis's visits is supported best by scientific data. The period is well defined and has very synchronous dating worldwide. A rapid return of glacial conditions in the higher latitudes of the North American Hemisphere occurred and is believed to have been created in a very short period of 10 to 20 years. Data from oxygen isotopes, dust concentration, and snow accumulation of ice cores indicate warming of 7 degrees Centigrade in just a few years with total warming of Greenland being about 10°C.[34] A sea level rise began to pulse during this time that led to 60 more meters of height before attaining present levels about 7500 years ago.[24] Measurements of oxygen isotopes from the Greenland ice core indicate the ending the Younger Dryas took place over just 40 to 50 years. Other data such as dust concentrations and snow accumulation suggest an even more rapid transition requiring about 7 degrees Centigrade warming in just a few years. The end of this period is determined with very high resolution of numerous sources from various disciplines to be 11,500 years BP.[34]

A currently, partially, accepted scientific explanation is the failure of the North Atlantic Ocean's thermohaline circulation due to sudden influx of fresh water from rapid glacial melting in North America. Other hypotheses are an impact event hidden because the impact location was on the ice cap, and unusually large global volcanic activity.[35] However, no one from the respected academic community has ventured to propose disturbances created by visiting celestial bodies on a grand scale such as the near passing of the Nemesis star. Numerous worldwide myths do account for celestial events during this time. More than likely, as this paper is proposing, the close approach of Nemesis with Earth brought red dust, comets, asteroids, and a combination of perturbing gravitational and electromagnetic forces. These forces possibly shifted and twisted the Earth's mantle and crust about its liquid core. This rotation of even 10 to 20 degrees would cause great distress on the Earth's crust because the oblateness of 6.5 miles on radius of the Earth's surface about the equator caused by centripetal force. The Earth's geoid would need to adjust globally. This re-adjusting of the Earth's crustal altitude would create some wobbling or shifting of the spin axis, possible reversal of the magnetic poles, horrendous storms, the deep freeze of flora and fauna that was suddenly moved into the new polar region, increased volcanism, earthquakes, the movement of partial or entire ice caps that can slide into the ocean, dramatic rise of sea level, and the complete drainage and/or re-location of inland lakes and parts of some seas.

These sweeping and superfast changes of Earth's surface would cause utter destruction of any of mankind's existing civilizations, especially of the concentrations of humanity located at seaports and along rivers. Many fauna categories are driven into extinction. However, there were enough survivors for life to go on, especially for man. Legends of this great destruction and flooding caused by celestial visitors were handed down from generation to generation. Many worldwide documented records (long thought to be legends with no technical basis) were eventually created such as the account of the Noah flood. The post Younger Dryas times are believed by archaeologists to cause drought in the Levant region or Middle East motivating surviving cultures to develop agriculture.[36] This hypothesis seems correct since that part of the world would have moved from less tropical latitude to drier northern latitude and avoided severe ocean coastal flooding by remaining on the Earth's more elevated, oblate crust. The prediction of this paper is that the Earth's crust rotated about 20° latitude moving Siberia and Eurasia that much closer to the North Pole. Canada and northern United States moved an equivalent amount away from the North Pole. Hence, a thick permafrost was built up in Siberia, and the existing ice cap melted in Canada and northern United States.[37] This combined crustal movement would not exclude the continent of Antarctica which moved directly over the South Pole. More "food for thought" is the conjecture that part of the mythical land of Mu was on the western part of this continent and is now buried under an ice cap.

Other evidence that supports this dramatic change of the Earth's surface toward the end of the Younger Dryas period is indisputably listed[34]:

- Replacement of forest in Scandinavia with glacial tundra

- Piles of mammoth bones and other indigenous animals found on islands in northern Siberia

- Extinct animals found freeze-dried (only caused by a very sudden lowering of temperature and continuing freezing temperatures) and even used as food by gold prospectors in Alaska and the Yukon

- Both prey and predator of extinct animals found in caves together seeking shelter from something more feared than being eaten

- Young wooly mammoth calf found freeze-dried with daisies in its mouth now exhibited in a natural history museum in St. Petersburg, Russia

- Mammoth and sabre tooth tigers' bones found in fishing nets that dragged the sea bottom 200 feet below sea level on the North Atlantic continental shelf

- Glaciation due to increased snow in mountain ranges around the world

- More dust in the atmosphere coming from deserts in Asia where rapid deforestation and high winds occurred

- The Huelmo/Mascardi Cold Reversal in the Southern Hemisphere ending at the same time; any possible pole shift should have caused symmetrical climatic changes about the equator.

- The American ice sheet of the last ice age was centered on southern Canada below the Hudson Bay while the current polar region of those times was free of ice suggesting the Earth's lithosphere had shifted 20 to 30 degrees in latitude which may explain the difference between the Earth's spin axis and its moving magnetic pole axis. Currently, the northern magnetic pole is slowly moving toward the spin axis pole; perhaps the residual magnetism within the Earth's crust and mantle from the previous pole location is making adjustments to seek the true magnetic north over the new spin axis location.

- Decline of the Clovis Culture and extinction of animal species in North America. One ridiculous conjecture is that the Clovis people over-hunted the 'wooly mammoth' with their primitive Neolithic weapons with no explanation as to why smaller animals also went extinct.

Clear explanations for all these events happening very close together is lacking unless there was a slight but dramatic pole shift caused by the god-like Nemesis.

XIII.

THE UNEASY MARRIAGE OF MYTH AND SCIENCE

Why was the Great Deluge or Flood a myth created by almost all surviving cultures around the world whose peoples had supposedly little or no contact with each other? Scientific and technical reasons are slowly coming to the forefront to support the myths. And the Sun's sister star, Nemesis, gives us another reason. Only until recent times have the discovery of such bodies as brown dwarfs and binary star systems with planets become known to astronomers. Is it truly possible that our very own Sun is part of a binary system? Did our star system in its early formation inside a star-burst cluster capture another star system that continues to orbit the more massive Sun in an elongated orbit and disturb the tranquility of the Sun's supposedly pristine planetary system? More news arrives daily from space probes and space telescopes that indicate the solar system is not so tranquil and constant.

Scientists must ever be vigilant to keep open minds and explore ideas that are outside the accepted norm. The previous discussions do certainly suggest more specific research such as:

1. Continuing the search for a nearby brown dwarf in the infrared spectrum if not already found.

2. Observing the Sun along with background stars looking for unusual wobble or weaving not attributed to its own known planets.

3. Performing more experiments with plasma to assure how various discharges should appear in space.

4. Postulating and modeling the magnetic field strengths of brown and red dwarf stars and how they may generate plasma discharges.

5. Demonstrate in the lab by scaling how plasma discharges of what power can cause the surface anomalies discovered on Mars.

It is time to gather other very interesting and unique viewpoints that corroborate the proposals of this paper. These views and predictions are mostly written by authors considered to be catastrophists who have ignored or defied uniformitarianism and the other assumed laws of nature and physics. The laws of nature are simply those ideas that are currently fashionable; new discoveries and ground-breaking ideas have continually destroyed the contention that any of man's laws of nature are immutable. The predictions of this paper are based on reasonable scientific data that can lead to troublesome times. Governments, religious bodies, and universities want to completely dismiss catastrophism with no or little fair review. These ideas cannot simply be dismissed as the products of doomsayers and crack-pot scientists. These kinds of predictions actually have guided curious and serious researchers into areas of climatic studies via ice and seabed coring, more focus on modeling the Earth's mantle and lithosphere, more studies of correlating rock dating on Moon and Mars, increased observations and tracking of near-Earth asteroids, exo-solar planetary searches, etc. Major contributors to this paper's hypothesis are listed and cited in the following table. Their predictions and ideas have significantly led to the conclusion that Nemesis exists and possibly threatens our spaceship, Earth. Although many of these authors' theories are controversial and/or completely dismissed by the scientific community and also by this paper, many other of their excellent ideas simply cannot be ignored and must be given our careful attention.

Table 23: Major Contributors to the Evolution of the Ideas for the New Nemesis

(This table by no means infers that the listed authors completely accept each other or this paper's hypothesis.)

Author	Approximate Publish Date	Book(s) or Journal Published and Discussed
Immanuel Velikovsky	1950	*Worlds in Collision*
Charles Hapgood	1958 / 1968	*Earth's Shifting Crust / The Path of the Pole*
W.P. Farrand	1961	"Frozen Mammoth and Modern Geology" by *Science Journal*
Eric von Daniken	1968	*Chariots of the Gods*

Author	Approximate Publish Date	Book(s) or Journal Published and Discussed
Zecharia Sitchin	1976	*The Twelfth Planet*
John White	1980	*Pole Shift*
David Talbott	1980	*The Saturn Myth*
D.S.Allan and J.B.Delair	1997	*Cataclysm! Compelling Evidence of a Cosmic Catastrophe in 9500 BC*
David Talbott and Wallace Thornhill	2005 / 2007	*Thunderbolt of the Gods / The Electric Universe*
David Talbott and Wallace Thornhill	2011	"Symbols of an Alien Sky"; "Episode 2: The Lightening Scarred Planet Mars"; "Episode 3: The Electric Comet" produced by DVD and YouTube movies and video streaming
David Talbott and Wallace Thornhill	2011	Thunderbolt Project: a series of video streaming documentaries.

XIV.

THE STORY OF A COSMIC CATASTROPHE 11,500 YEARS AGO

The first book of extreme interest for this paper is *Cataclysm! Compelling Evidence of a Cosmic Catastrophe in 9500 B.C.* by D. S. Allan and J. B. Delair.[38] The authors are a Cambridge M.A. science historian specializing in paleogeography and a B.Sc. Oxford-based geologist and anthropologist who is museum curator of geology at the University of Southampton, England. This multi-disciplinary study examines the great global catastrophe that occurred 11,500 years ago. The book models a cosmic intruder, the resulting displaced spin axis, geological consequences; the resulting extinctions of flora and fauna; and the worldwide legends and traditions that portray this terrifying event.

These authors dubbed the celestial body that brought hell on Earth as "Phaeton" the French form of "Phaethon" which is the Latin form given by the Roman poet, Ovid. As the myth states: Phaeton, who is the son of a solar deity, the Sun, gained permission to ride the Sun's chariot. Phaeton was unable to control the chariot almost destroying Earth; Zeus (of Greek origin) or Jupiter (the Roman version of Zeus) killed him with a thunderbolt to save Earth or Gaia from total destruction. This documented myth was passed from culture to culture via re-interpretations of different ancient languages throughout the millennia of mankind's known existence: to the Romans who called this deity Phaeton from the Greeks (whose name for Phaeton is Typhon) that was found in Homer's writing and Hesiod's *Theogony* and *Works and Days*[39]; to the Greeks from the Assyrians (whose name for Typhon is Ashur, depicted as a god within a winged disc) that was taken from the tablets of their creation epic found in Nineveh[40]; to the Assyrians from the Babylonians (whose name for Ashur is Marduk) taken from the writing of Enuma Elish, meaning "when in the heights"[41]); to the Assyrians from the Akkadians (whose name for Marduk is Erra), taken from their text of Genesis written in Old Babylonian dialect[42]; to the Akkadians from the Sumerians (whose name for Erra is Nibiru or Elohim, the spirit) found in their Epic Tale of Creation found on seven clay cuneiform tablets, with six dealing with the creation process. The symbolic depiction for Nibiru is the winged disc.[43] The people of Sumer claim their epic tale was handed down from previous millennia.

Table 24: Lineage of Nemesis's Name

Evolution of Nemesis's Name	Cultural Origin	Literary or Written Source or Idea Attributed to the Nemesis Name
Nibiru or Elohim	Sumerian	Epic Tale of Creation created on clay cuneiform tablets.
Erra	Akkadians	Text of Genesis written in Old Babylonian dialect.
Marduk (possibly means the planet of Nemesis)	Babylonians	Writings of Enuma Elish, meaning "when in the heights".
Ashur	Assyrians	Tablets of their creation epic found in the city of Nineveh.
Typhon	Greeks	Homer's and Hesiod's writings in the *Theogony* and *Works and Days*.
Phaeton or Phaethon	Roman	Writings of a Roman poet, Ovid
Planet "X"	Modern times	A search beginning in the 20th century for unknown planet(s) was initiated to explain orbital perturbations of the outer planets.
Nemesis of Myth	Ancient Roman and Greek times	In Greek mythology, she is the inescapable goddess of revenge and in Roman, the avenger of crime.

Evolution of Nemesis's Name	Cultural Origin	Literary or Written Source or Idea Attributed to the Nemesis Name
Nemesis of Mass Extinctions	Very recent times	A hypothetical red dwarf or brown dwarf star was postulated in 1984 to be orbiting the Sun at a distance of about 95,000 AU, somewhat beyond the Oort cloud, to explain a perceived cycle of mass extinctions in the geological record.
Nemesis of the Saros cycle returning approximately every 3600 years	Currently proposed by this paper and partially by author, Zecharia Sitchin, in 1976	This hypothetical orbiting red or brown dwarf star is responsible for a shorter cycle of catastrophic events some of which are recorded by man.

The Middle Eastern religions basically shared the same pantheon and belief system. For the earliest of cultures these so-called myths were the merger of their religion and science. These myths were eventually passed to the Hebrews who combined all of Sumer's Anunnaki deities in their Old Testament into one deity who lived in the heavens, on Earth, and in spirit; but the Hebrews still retained the original Genesis story in a very condensed version. This idea was then passed to Christian and Muslim religions that retained most of the Hebrew's Old Testament. The authors of the *Cataclysm!* are suggesting that the Phaeton myth is actually symbolic of a real earthly event that occurred and was handed down through the ages from culture to culture.

They backup their belief by citing and cataloging over 100 traditions or myths or legends from ancient indigenous peoples from every continent (except Antarctica) that specify very similar catastrophic effects. The common themes of these effects are conflagration prior to flooding, global non-riverine flooding, disorder of objects in the sky, terrestrial chaos from volcanoes and earthquakes, prolonged darkness, falls of hail, dust and fiery objects, and subsequent colder climates. I state verbatim the conclusion of these authors which any present scholar or scientist can hardly deny, "These same memories, when assessed collectively, consistently appear to describe a most amazing episode of Earth history upon which orthodoxy, represented by astronomy, geology, and archeology, has so far remained largely silent. It is, in fact, this internal consistency of these memories which – irrespective of the original geographical or cultural source of the material – is so impressive. .... they constitute an eloquent testimony to a truly momentous chapter of events."[38]

Their conclusion is extremely forceful when added to their compilations of enormous topographical changes and terrible biological extinctions and decimations that are largely dated around 11,500 years ago. How can such a book be largely overlooked by the scientific community and the current media? Velikovsky, who was both a psychoanalyst and historian, predicted that such a horrible disaster can be blotted from human memory by continual collective amnesia similar to individuals that submerge painful experiences from normal recall.

XV.

CRITICISM AND QUESTIONS FOR AUTHORS ALLAN AND DELAIR

Their book challenges scientific dogma and receives mostly silence and sequestration from the combined academic, political, and religious communities. The book can be definitely criticized on scientific grounds by its huge difference between the accepted theory and these authors' ideas. They do not accept the Ice Age theory prior to the Younger Dryas Period and the reasons for the northern ice cap receding. Many of their ideas come from a quickly shifting of Earth's poles which is totally rejected by modern science. A full discussion of shifting poles will soon follow in this paper. The book is foggy on whether total pole shift or some crustal shift occurred and what mechanism triggered such an event. However, a compelling argument for reversed polarity of the poles occurring during this event is given; strong electromagnetic forces caused by Phaeton's close encounter with Earth are given as the trigger for this occurrence. I can definitely agree with this part of their hypothesis; however, this idea still is considered highly variant from normal dogma.

And, *Cataclysm!* does not model very well the celestial disturbances that triggered revelation on Earth. There is no mention of a visiting Nemesis star which is the chief cause for Earth's calamitous history per this paper; Allan and Delair placed all the blame on the mythical planet, Phaeton, whose path is diagramed inside their book. Phaeton is postulated to perhaps have come from the Vela supernova dated about the same time as a highly electromagnetic formed body. The diagram reveals that Phaeton takes one path through the solar system causing Neptune to lose Pluto and derange its other moons; causing equatorial misalignment of Uranus; displacing Saturn's moon, Chiron; breaking up the planet Tiamat and detaching its moon, Kingu, becoming Earth's Moon; disturbing the surface of Mars and changing its orbit; having a close encounter with Earth; changing the spin of Venus, and then finally crashing into the Sun.[44] This actually comical path of Phaeton is complete nonsense and is laughable in the astrophysics and planetary science disciplines. Also, astrophysicists definitely dispute that fast-moving celestial bodies exterior to the fast-moving solar system can actually be captured due to the laws of physics and calculus. The authors are attempting to make sense of combining the literature and traditions of the Hellenic authors, ancient religious sources, Egyptian and Persian depictions and epics, and the cuneiform tablets of the Akkadians and Sumerians. This blending is ludicrous since so much of the historical data that is used becomes distorted in re-interpretations down through the ages. Perhaps some or all these events might have occurred with Nemesis but certainly not with only one pass of Phaeton through the solar system; recurring passes of Phaeton are required to make any of these events plausible. The ancient literature will certainly reveal trends but will utterly fail to correctly detail any changes in orbital paths or close encounters of the planets by solar system intruders. The timeline for all these celestial events is truly only guesswork, although a listing of actual events occurring on Earth via myths and geological/biological data can be roughly constructed.

I have definite criticisms for Allan and Delair, but this does not disqualify the majority of their very detailed and well-organized presentation. Unfortunately, one or more glaring mistakes are cause for many academicians to completely dismiss the book. Most scholars prefer to devote their entire career on small pieces of knowledge moving understanding ahead in smaller, carefully researched steps. Scholars and the scientific method scowl at people who take such large, sometimes reckless steps forward such as Allan and Delair. However, the information and ideas presented in their book are indispensable to understanding the complete history of mankind coming forward from 20,000 thousand years ago.

XVI.

SHIFTING OF EARTH'S CRUST AND/OR SPIN AXIS

Another book for our review that was published in 1980 is *Pole Shift* by John White.[45] There are numerous publications that have placed predictions of past and future pole shifts, but White's book definitely placed this idea on the public radar.

Ideas that preceded White are from *Earth's Shifting Crust* and *The Path of the Pole* by Charles H. Hapgood published in 1958 and 1968 respectively.[46] The main driver of these books is an article published in the *Science* journal in 1961 by W.P. Farrand called "Frozen Mammoth and Modern Geology".[47] This article carefully described that wooly mammoth bones were piled into small mountains on islands north of Siberia and in certain places in Alaska. Many carcasses were preserved by quickly being frozen; one young calf had daisies in its mouth. Allan and Delair chronicled piles of bones of both predator and prey found deep in caves that came from the same period. The data that was collected from these extinct species certainly shows something very dramatic on Earth's surface occurred. Such explanations of a comet strike or a rapid change in the North Atlantic currents does not explain such a dramatic change in climate or weather that would almost rapidly freeze-dry large pasturing animals or drive both prey and predators into caves for shelter.

Hapgood, referenced by White, suggested the Earth's surface crust or lithosphere shifted slowly over years instead of hours as White proposes. Neither author dismisses continental drift and sea floor spreading (plate tectonics) as existing conditions, but they are considered secondary to more treacherous global crustal shift. Hapgood originally postulated the ice caps dramatically slipped causing crustal shift, and then modified his trigger to be a "wave guide layer" or gravitationally unstable liquid layer in the asthenosphere about 100 miles below the surface. The forces required to move a sizable unified portion of a 100-mile-thick crust with respect to the mantle assuming higher viscosity than liquids for the top of the mantle simply cannot be explained by classical physics. Hapgood did extract some important evidence that prior to the last glaciation in North America the pole stood in Hudson Bay and shifted to its present site in a gradual motion that began 18,000 to 17,000 thousand years ago and was completed by about 12,000 years ago. The pole at Hudson Bay arrived from the Greenland Sea about 50,000 years ago. Hapgood came amazingly close to what possibly occurred near the end of the Younger Dryas Period but was reluctant to disagree with the present fashion of uniformitarianism and suggest a more rapid shift. Hapgood lacked a scientifically acceptable cause or trigger for shifting the crust.

Then White stepped into the fray to suggest a more dramatic shift where the entire Earth mass is shifted on its spin axis within days or hours. More data collected about the mass extinctions occurring after the Younger Dryas Period convinced him that catastrophism is closer to the truth. White needed a plausible cause and suggested a cosmic disturbance taken from some of Velikovsky's ideas. White came up against classical physics in trying to explain how the tremendous spinning mass of the Earth can change its immense angular and kinetic momentum. One proposal is that the entire mass of the Earth flipped 180°. No amount of gravitational force from the close encounter of a celestial body can possibly transfer that required energy without heating the Earth to red hot temperatures. White's possible confusion came from trying to explain geomagnetic reversals which were recently discovered during his time. Heliomagnetic reversals occur every 11 years on the Sun, but the Sun does not flip. Planetary scientists still consider celestial body magnetic reversals as a mystery, but certainly do not attribute them to the entire planet flipping on its axis.

An epilogue topic (pp. 395 to 399) of *Pole Shift* focuses on an interesting debate about frozen mammoths. Two scholarly correspondents of John White, William White (an opponent of flash-freeze school of thinking) and Dwardu Cardona, argue about how these animals died - whether by asphyxia before freezing or by freezing. Then comparisons of the preserved frozen bodies verses putrefaction were presented. Cardona discusses a certain Berezovka mammoth that was found, "The position in which this beast was found clearly indicates that it could neither have been drowned nor crushed beneath a slide. Its stance suggests that it was felled on its haunches, that it attempted to regain its feet, that it was then somehow asphyxiated, and that it froze in this animated position. It did not even keel over." This detailed description perfectly captures the picture for catastrophism and begs to have these kinds of recorded events resolved.

White brings to the forefront that Earth's geological and biological record in the recent past of 20,000 years demands careful scrutiny and some type of deductive reasoning for its modeling. White brings forward the Edgar Cayce, Nostradamus, Biblical, and other prophesies in this book and immediately alienates the scientific community. But his question and attempted answer rings out loud and clear – what happened 11,500 years ago?

XVII.
VELIKOVSKY TRIES TO EXPLAIN
MYTHS IN TECHNICAL TERMS

The next book for review outraged the scientific community when it was published in 1950 by MacMillan. Immanuel Velikovsky's *Worlds in Collision* seriously challenged scientific dogma.[48] MacMillan was forced to cancel publication because it produced textbooks for universities that immediately pressured the publisher. Velikovsky was the victim of intellectual arrogance, fragile assumptions, and cliquish associations. He reconstructed history from long ignored data found in myths and ancient literature that he corroborated with the physical, biological and social sciences.

Velikovsky, by getting some ideas from the interpretation of ancient texts including the Jewish Old Testament, proposed that Jupiter underwent a shattering convulsion that produced the planet, Venus, which in turn menaced Earth and Mars. The storyline is faintly familiar with Allan and Delair's attempt to interpret ancient text, but Jupiter is used this time instead of Phaeton as the culprit. As mentioned previously, interpreting ancient texts can lead to similar lines of thought but the interpretations can quickly diverge and become meaningless when being explained by scientific methods in classical physics. The interpretations are intriguing but are sometimes just fanciful guesswork and have no underpinning of celestial mechanics.

Another theory of Velikovsky's was that the Earth turned part-way over on its axis, as supposedly was written in Persian and Chinese ancient texts, and introduced the idea of electromagnetic forces or charges occurring between Venus, Mars, and Earth with their close encounters or with other celestial bodies. Again, Newtonian physics could not support this belief, although the author had connected the idea of planetary magnetic reversals with some of these myths. The cause of magnetic reversals is not known, but certainly do not represent the flipping of Earth on it axis. Velikovsky wandered too far from scientific understanding of his day, but did rejuvenate the modern thought process for the re-marriage of myth and science.

XVIII.

THE CONNECTION BETWEEN MYTHS AND UNKNOWN PLANETS BY SITCHIN

Another book that continued this marriage was the *Twelfth Planet* by Zecharia Sitchin first published in 1976.[49] Sitchin's unusual ideas are based on supposedly accurate interpretations of ancient texts of the Babylonians, Akkadians, and Sumerians. He amazingly addressed postulations and reasons for the Great Deluge, the Earth-Moon system, the asteroid belt, the observed and recorded twelve wanderers of the sky that included Nemesis, ancient mathematics and the genetic creation of mankind in the image of ancient astronauts. All this knowledge came from texts describing very intelligent beings that landed on Earth about 250,000 years ago from a planet that passes through our part of the solar system every 3600 years. The planet is supposedly part of the Nemesis star system. His ideas were too bizarre and all his books were condemned to the occult section of libraries. Surprisingly, only until very recently are his interpretations of ancient texts debunked by experts in the field of cuneiform, hieroglyphics, and other ancient languages. What took so long? Simple denial of ancient man's myths having any scientific connection remains steadfast. His ideas do lend support to why and how ancient monoliths such as the Giza's Great Pyramids or monolith structures of the Sacred Valley in Peru could have been built with unknown, superior technology by these more highly evolved visitors from another celestial body.

Of course, his strange model of the past is totally rejected for numerous reasons; some of the more important rejections are listed along with some plausibility for this ancient astronaut model.

1. There is no evidence of tools or materials of a past superior technology, except possibly for the unexplained and amazing, consistent monolith structures that are globally found around the planet.

2. If these celestial visitors were superior, why did they disappear? Perhaps the event of the Great Deluge destroyed them as well as most of mankind. A small

vestige of aliens or alien-made humans survived long enough to pass on the stories told in ancient texts.

3. There are no astronomical observations of any nearby star or planets orbiting the Sun. Of course, if Nemesis is a dark brown dwarf star that is way beyond Pluto for most of its orbit, then its sighting even with modern telescopes would be difficult.

4. No living beings similar to man can survive on a cold planet orbiting a brown dwarf and not receiving sufficient radiant energy from the Sun or another body for most of its orbit. Possibly, the planet's internal heat and infrared heat from the brown dwarf provide the necessary radiation, and light comes from broad bands of atmospheric auroras similar to what is observed on some of the Sun's planets but is more dramatic and radiant. Recently postulated by a group of independent scientists, called the "Electric Universe" group, is the idea that a plasma sheath normally surrounds a brown or red dwarf star. This protected dim plasma glow sheath, which is considered a bloated corona of the star, engulfs its orbiting planets. The planets, which receive the star's radiant energy, light and also water, can supposedly support life as we know it, using a similar light spectrum and temperature range as exists on Earth.

5. The repeated chaos caused by the intersection of two planetary systems every 3600 years leads to question how an intelligent, civilized set of beings could have had enough uninterrupted time to evolve on a planet orbiting Nemesis. For me, this is actually the predominant reason for Sitchin's model being questionable and not occurring. Of course, the modeling of this type of system via modern celestial mechanics along with probability and statistics could prove me wrong. Various dinosaur species survived indisputably for millions of years without interruption. One may want to wander more to the "wild side" and venture a guess that the aliens had enough uninterrupted time to evolve enough to understand and utilize space travel between planets. These aliens could predict which planet of which star system would be more endangered and travel to another planet for maximum safety. This type of planned inter-planetary transfer might have then occurred for thousands and thousands of years. Another reasonable version is that humanity progressed one or more times to great technical feats without the intervention of extraterrestrials, and then was repeatedly destroyed through catastrophic events

6. Until recently, since the discovery of exo-solar planets, the scientific dogma rejected any conjecture that planets can normally exist around red or brown dwarf stars due to the nebular hypothesis' tenants. So, the model that Sitchin predicts about a planetary system being captured and orbiting the Sun is quite possible and is now acceptable in scientific terms.

7. Increasing knowledge of genetics and the missing link for mankind's evolution does not make man's birth from alien astronauts' laboratories so fanciful. Also, another species of humanoid with elongated skulls discovered worldwide could be likely evidence of an alien species or an earlier species produced by the aliens.

8. What is amazing is why Sitchin was not attacked until very recently on the grounds of the study of archeology and the study of ancient languages. Apparently, Sitchin's more important translations and interpretations are considered to be largely in error and have been debunked – perhaps by some jealous or well-paid charlatan. There only exist a few hundred specialists worldwide in the field of ancient languages that really have an opportunity to speak-out against or for Sitchin. Meanwhile, how and why the worldwide pyramids and other monolith structures requiring unknown technology were built remains very much a mystery.

XIX.

SITCHIN PARTIALLY REDEEMED BY THE CORROBORATION OF NASA DATA

One of Sitchin's interpretations discussing the reasons for Tiamat, the watery planet; the "hammered bracelet" or asteroid belt; and the Earth-Moon system are partially applied to one of the hypotheses presented in www.ettingerjournals.com. The journal called the "Earth's Metamorphosis Hypothesis" explains with classical physics and calculations the genesis of the Earth-Moon system.[50] This new genesis story disagrees with NASA's currently accepted model, but incredibly aligns itself with the dating of Earth and Moon rocks and with a description of the Genesis of the Jewish Old Testament as interpreted by the Gideon Holy Bible. The ideas of this Earth-metamorphosis model align with much of the celestial events described in the *Twelfth Planet*. The re-marriage of myth and science continues. Can the scientific community afford to keep totally rejecting consistent worldwide myths as possible evidence of past, true events?

The Gideon Holy Bible uses more modern language taken from the St. James version of the Christian bible. One word in the Genesis story, "firmament"[51] is largely misunderstood by both Sitchin and Velikovsky. Both authors believe the "firmament" exists in the celestial sky and in the outer solar system due to its connection with "heaven". This word is also discussed in the legends about the goddess or planet called 'Tiamat' where the waters are divided by the "firmament". This paper has another hypothesis for the "firmament". The division of certain types of waters is discussed in the previously referenced "Earth's Metamorphosis Hypothesis". A rogue icy planet strikes and penetrates the watery surface of an already differentiated planet called 'Tiamat' (what becomes Earth); a tremendous outflowing of the Earth's mantle from the impact crater flows over the surface to create the "firmament" that is called heaven and separates the waters; these waters are the water that originally covered 'Tiamat' or Earth and the melted waters and other volatiles of the rogue planet that are trapped inside Earth's mantle under the newly raised dry lands and original oceanic crust. In other words, some of Sitchin's model has supposedly been corrected and redeemed not only to account for a better translation of the ancient texts, but also to account for recently gathered geological evidence from both the Moon and Earth.

Consensus science only considers gravitational and kinetic reasons for how celestial motions are determined. Presently, the scientific community rule out that electrical and magnetic forces on a star-system scale actually dominate over gravitational forces. If one simulates the solar system on a computer using only Newtonian concepts, perturbations between the planets and their moons eventually become disturbed enough to cause chaos in the system. NASA has no answer except to say that this puzzlement is yet to be solved. The only answer is that other forces, not addressed, are maintaining the system. The Sun with its magneto-sheath and solar winds, the planets with their magneto-sheaths and general changing charge unbalances act like electrical circuitry to restore the orbital radii if they are perturbed such as would happen when Nemesis and its planets come close to the planetary orbits of the Sun. The independent scientists of the ElectricUniverse.com (EU) claim that there is an unbalance of charge occurring between the planets and between the planets and the Sun with its positive charge. A massive electrical circuitry and equilibrium of charge is always working to maintain certain distances between planetary bodies and the Sun. The contention of this paper is that after Nemesis cruises through the Sun's magneto-sheath and its planetary system, orbital perturbations do occur, but are soon returned to normal after Nemesis leaves. Ample time is available for effected electric and magnetic fields to reach equilibrium once again and preserve the orbital shapes and distances of the various planets. Of course, very infrequently during a crossing of Nemesis a close encounter and/or violent exchange(s) of plasma energy will occur causing calamity.

The study of close multi-star systems and exo-solar systems and 'hot Jupiters' well inside Mercury's orbital distance from stars reveals that stability does exist under extreme conditions when gravitational forces fail to explain their genesis or continuance. Electrical forces are controlling these celestial bodies on a macro-scale and this is why astronomers can observe so many at any one point in time. These same electrical forces control the particles inside matter on a micro-scale. Orbiting systems, no matter what their scale are kept stable by electrical forces that unify the entire universe. Gravity is a minor actor except for mankind's experiences on Earth and with his space probes in space.

In conclusion, a star system with all its attendant celestial bodies is controlled significantly more by electromagnetic forces than by gravitational forces. This concept is why an elongated orbit of a Nemesis star around the Sun can be stable for millions of years, and possibly the life of the solar system, discounting some infrequent incursions of close encounters. These close encounters, and possibly some collisions, result in calamity that quickly subsides and is forgotten. Only certain surface features on these celestial bodies and Earth's climate data show a record of these violent visitations.

The Nemesis star is known by many names through the ages. The Sumerians called it Nibiru, the Akkadians called it Erra, the Babylonians as Marduk, the Greeks as Typhon, the Romans as Phaeton, and in modern times as Planet "X" or as Nemesis, the name given by NASA. Of course, not all the above cultures witnessed Nemesis. Many intervening cultures received stories about Nemesis through legends and traditions that modern historians believe to be largely embellishments and metaphors. These stories are indeed just that, but also true. This second star in mankind's sky is considered as a proto-planet, Saturn, for the Electric Universe people that promote the Saturn polar configuration model. Supposedly, Saturn is a captured brown dwarf having Earth, Mars, and Venus as planets which were then re-configured into orbits around the Sun. If it sounds too confusing, refer to Talbott's book, *The Saturn Myth*. Talbott, being a comparative mythologist like Immanuel Velikovsky, attempts to translate and transform the ancient myths into a real storyline that produces a very different solar system from the current one. However, embellishments of various translations from one language to another by different cultures spanning several millennia can lead one astray as I proposed happened with Velikovsky and Talbott. A picture as Talbott imagines the polar configuration being viewed from Earth follows. This archetype is labeled the Cosmic Wheel with a tongue or mountain. The claim of this book is that the archetype did appear during the Younger Dryas, but represents the plasma discharge of the alignment of Nemesis and some planets over the Earth's pre-deluge North Pole.

Nemesis recently returned to Earth's skies in 5900 BC (Sumerian), 2300 BC (Egyptian), and 1300 AD (Medieval) which is in line with actual witnesses that revered or feared and recorded these observations for the flourishing cultures of Mesopotamia (Sumerian and Babylonian) and Egypt during two of these cited times. These cycles were first identified by Zecharia Sitchin in his book, *The Twelfth Planet*, as Sar cycles. The cycles were then identified in my papers by pinning the most likely return to 9500 BC when the Younger Dryas and Holocene extinction event were unquestionably dated. Then, I added or subtracted 3600 years from that point in time to tract other visits by Nemesis.

XX.

THE CONNECTION OF ANCIENT ROCK ART TO THE NEMESIS STAR

This paper now discusses another amazing and important juncture in the re-marriage of science and myth. David N. Talbott co-authored two books with Wallace Thornhill, *Thunderbolts of the Gods* in 2005 and *The Electric Universe* in 2007, which have interpretations of rock art found worldwide that merges with the newest findings from space probes to Mars.[52] Talbott also maintains an active website called www.thunderbolts.info that offers an excellent display of ancient artifacts and knowledge of myths. His convincing oratorical skills are very enjoyable and have obviously blended his comparative mythology with Wallace Thornhill's knowledge of the science of plasma and recent probe missions to Mars and to comets.

David Talbott originally promoted neo-Velikovskian concepts and wrote a book in 1980, *The Saturn Myth*.[53] He made the same mistake as Velikovsky in trying to interpret too much from human memory and ancient artifacts. He attempted to describe an earlier solar system which is in flagrant disagreement with celestial mechanics and has no possible chance for explaining the evolution of flora and fauna on Earth. He proposed a "Polar Configuration" involving four planets, Saturn, Venus, Mars, and Earth in that order which formerly orbited the Sun as a linear assembly while it rotated about its barycenter and influenced human mythology. A convincing quote from Talbott regarding myths and ancient traditions is, "How did human consciousness, emerging from the womb of nature, converge on the same improbable ideas by 'contradicting nature'"? Obviously, Talbott also wished to contradict nature, too, with his "Polar Configuration" which supposedly portrays the true meaning of his study of worldwide myths. His first marriage with science failed but his persuasive comparative mythologies live on.

Talbott years later merged some of his ideas with physicists Wallace Thornhill and Anthony Peratt, to explain a natural phenomenon of plasma displays in the ancient celestial sky depicted by both ancient rock art and their similarities with laboratory experiments of plasma discharges. Talbott is now having a more successful second marriage with science with his latest books and recent website. Viewing his website actually inspired this paper. These intriguing ideas of Talbott still lack a model that can make sense in the field of celestial mechanics and the geological and archaeological history of Earth. My own concepts attempt to derive a reasonable model with certain limitations that can be accepted by mythologists, archeologists, astrophysicists and planetary scientists.

In his website article "Symbols of an Alien Sky" Talbott describes the different rock art depictions and how they are very consistent worldwide. He claims that ancient peoples were drawing mostly about events that they saw in the sky.[54] Their art was a carnival of ghostly creatures with absurd patterns. Some of the most common archetypes were stick men with different heads and squatter men with twin dots on either side (found both in Egypt and Arizona). Thornhill and Peratt recognized many of these forms from experiments performed in plasma labs, especially the toroidal synchrotron radiation represented by the two dots or circles on each side of the squatter man. The upraised multi-arms and spread legs of the stick/squatter men represent squashed bells of sheet radiation caused by diocotron instability. The vertical line or body of the stick men is the conduit of charge called Birkeland currents traveling between two opposite magnetic poles. The head of the stick men, which is shown in many shapes and motifs (most common were birdlike and toroidal with two eyes), is supposedly the source of the radiation discharge such as could come from a highly magnetized region of the upper atmosphere.

Another archetype is the cosmic wheel or wheel throne whose inspiration and explanation per Talbott did not come from the Sun. The spokes can be of many varieties but are considered to have no function for supporting the rim of an actual wheel. The wheel is many times connected to other lesser wheels and is sometimes shown with a crescent at different positions. The wheels are believed to represent alien celestial bodies with plasma discharge arcing or streamers reaching outward being either straight or wavelike. The crescents sometimes shown in different positions about the cosmic wheel possibly reveal phases of light cast by the Sun onto the surface of some strange large celestial body or the corona of a dwarf star. The most prominent and largest cosmic wheel is thought to be the planet Saturn per Talbott's interpretation, and probably appeared to ancient people as being the largest - larger than perhaps the angular diameter of the Moon in the sky. Celestial mechanics tells any astronomer that this is not possible; planets simply cannot rearrange themselves from orbiting Saturn to orbiting the Sun. This paper is not disputing that this archetype did appear in the sky, but is proposing the very likely probability that this cosmic wheel archetype represents one or more of the visitations of the Sun's sister star, Nemesis.

Let's compute a very possible angular diameter for Nemesis, as it appeared in the sky to our ancestors. A recently discovered brown dwarf, Teide 1, is estimated to be 1/10th of the Sun's radius.[8] The Sun is on average 1 AU from Earth whereas Nemesis which passes through the Main Belt of asteroids is 2.5 AU from the Sun and about (2.5 − 1.0) = 1.5 AU from the Earth. The Sun's angular size is on average 32.0 arc minutes. Then Nemesis's angular diameter is 32/10/1.5 = 2.13 arc minutes which is (2.13 x 60 arc seconds) / (Jupiter's 40 arc seconds[55]) = 3.2 times the size of Jupiter's image. Nemesis's apparent size in the sky can be significant; along with its radial discharge of columns of Birkeland currents and a possibly larger corona, it could easily look like a cosmic wheel in the sky possibly much larger than 2.13 arc minutes. If either Mars or Jupiter were in their shortest separation and alignment with Nemesis, then from the Earth this cosmic wheel will appear quite peculiar and spectacular to observers. As the alignments decreased and separation distance increased, the discharges could presumably become unstable and wavy as is also described by the ancients.

Another important and very common archetype is the cosmic thunderbolt believed to be major columns of Birkeland currents or plasma discharges emanating between celestial bodies – just as occurs today between the Sun and its planets with a corona ejection but only with a tiny fraction of the induced electrical current. The thunderbolt is also represented as a twisted filament which happens in the laboratory when the plasma arc becomes weak and starts to breakup. The thunderbolt was more likely a discharge column of plasma or bright Birkeland currents reaching between two celestial bodies as they came within close conjunction with each other, and this could be seen by man on Earth. Of course, this paper is introducing Nemesis as one of the new actors in this storyline. It is conjectured that possibly Nemesis, a very magnetic brown dwarf, passed between Mars and Jupiter and released a thunderbolt(s) toward these planets – possibly even as far as Saturn. It possibly appeared to the ancient sky gazers that their known planets were fighting with an intruder using a sword, trident, or thunderbolt. Talbott hypothesizes that the Gods of Mars, Jupiter, Saturn and Neptune are depicted as carrying thunderbolts or other weapons for this reason. These thunderbolts or electrical discharges may explain the interaction with the Nemesis star as seen on Earth at roughly 90 degrees from the path of columns of Birkeland currents.

Other archetypes found in ancient stone reliefs and tablets of ancient cultures are the radiant crown or magic helmet, the Pillar of Heaven, the cosmic wheel with a tongue, the side lock of the warrior king (the weakening of Birkeland currents due to the displacement of the source charge), the fiery serpent or dragon with streaming feathers (disturbed asteroids traveling through intense solar winds), the enclosed or spiraling serpent of creation and the Celtic cross. Many of these ancient archetypes can be represented by plasma technology in the laboratory. Hence, these archetypes are believed to have been observed in the sky as varying plasma displays between and around very electrical and magnetic ancient celestial bodies that have long disappeared. These bodies or planets were given names and believed to be gods that were revered and feared for they could and did bring chaos to Earth's inhabitants.

The comparative mythology of Talbott becomes scientific when simple inductive reasoning is applied. Why are these archetypes represented by Mesolithic rock art and stone friezes and clay tablet reliefs of ancient worldwide civilizations of the Neolithic era so consistent? Obviously, people globally from over 250 different indigenous tribes and civilizations saw the same events which can only be seen conclusively from the sky. These archetypes can only represent celestial events that were eventually given the names of gods and goddesses by the peoples of the earliest recorded civilizations which were passed on to subsequent cultures such as the Egyptians, Greeks, Hebrews, and Romans. Most of these gods and goddesses were the embodiment of planets or other celestial bodies existing in that one great common cinema in the sky.

Some the archetypes of the Thunderbolt Project are listed and re-interpreted by this author based on the Nemesis brown-dwarf hypothesis.

COSMIC WHEEL AS THE BROWN DWARF STAR, CALLED NEMESIS

The Cosmic Wheel has numerous versions but is mostly related to the brown or red dwarf star seen overhead during its crossing. When shown with the dark eye in the center, either Mars or one of its planets is eclipsing the Nemesis star from Earth's viewpoint. When the Cosmic Wheel has straight stable spokes, its anodic corona is interacting with some close encounter with another celestial body or the dwarf star is energized by highly-activated solar flares and winds. Strongly emitted plasma from Nemesis is depicted by straight spokes without a wheel rim. Strongly-received plasma from the Sun is depicted by spokes with a wheel rim and crescents on the corona surface. The crescents vary in position due to the position of the Nemesis star with respect to Earth's position with the Sun. These more illuminated crescents are either the Sun's light reflecting off the corona's envelope or the incoming solar wind current energizing the corona's surface. As the spokes of the Cosmic Wheel become wavy and break-up, the transmitted Birkeland currents are lessening due to increasing distances between the interacting bodies.

Typical Archetypes:		Typical Archetypes:	
Squatter Man		Cosmic Wheel with Wavy Spokes	
Stick Man with Raised Arms		Cosmic Wheel with Tongue	
Thunderbolt		Winged Disk	
Trident		Radiant Crowns	
Spiral Serpent		Hand of God with Eye	

These diagrams and interpretations of archetypes are taken from Thunderbolt.info by David Talbott.

Figure 25: Archetypes of Rock Art and Ancient Wall Reliefs and Friezes of the Ancient World

Cosmic Wheel with Tongue Occurs as a Dark Planet Passes Close to Earth

This paper postulates that one of Nemesis's largest planets, call it the Dark Planet, had a close encounter with the Earth's northern hemisphere in its normal, elliptical, orbital trajectory around Nemesis. This Dark Planet was either a highly-charged anode or cathode with its own strong magnetic field. Nemesis's planets share these characteristics by receiving and storing charge from the brown dwarf's anodic corona via interactions with the star's strong magnetic field. The close encounter with Earth created an illuminated, highly-charged Birkeland current column that completed an electrical circuit between the planets. The appearance of this highly-energized Dark Planet created the archetype of a *Cosmic Wheel with Tongue* when columns of Birkeland current were initiated and struck Earth.

Cosmic Wheel with Steep Mountain - Evolution of Energetic Birkeland Currents Reaching Earth

As the Birkeland currents increased their current flow and number of columns, the Cosmic Wheel appearance changed to a *High, Steep Mountain Having Various Radiant Crowns*. As the Dark Planet moves past Earth, the connecting Birkeland currents change appearance to a *Curved Side Lock*. As the Dark Planet moves farther away from Earth, the Birkeland currents appeared as *Half of a Radiant Star*. And finally, as the separation distance keeps increasing, the Birkeland currents are stretched enough to produce z-pinches. When the z-pinches occurred, the currents are still energetic enough to produce illuminations that look like the *Fiery Serpents/Dragons with Long Scales or Feathers* by viewers from Earth.

Winged Disk

The *Winged Disk,* as depicted by the Sumerians and other succeeding civilizations, has to represent a highly-activated Nemesis star passing through the Main Belt of asteroids at its perihelion with the Sun. The star electrifies the asteroids and other dust that illuminates their surfaces including the fine dust that sputters off their surfaces. From Earth this illumination of charged materials looks like wings emanating from the Nemesis disk or corona envelope. The disk should appear to Earthlings as several times larger than Jupiter at this distance.

It is possible that the largest planet orbiting Nemesis could also appear as a *Winged Disk* orbiting through Nemesis's corona when it is in close alignment with Earth. Another possibility is that Nemesis grows these wings as it crosses the helio-magnetosphere, thereby interacting with its double layer of current and the solar winds within the Sun's heliosphere.

Hand of God with Eye

I foretell two possible versions with the second one being the most likely for the Hand-of-God archetype. The first version is Mars being in line of sight between Earth and Nemesis. Nemesis is transferring highly-energetic illuminated plasma to Mars with a corona created around Mars as being the palm of the hand. The dark eye is Mars. The fingers are Birkeland column currents coming from Nemesis, which is being eclipsed.

In the second version Nemesis is transferring illuminated plasma to either Jupiter or Saturn, which is being eclipsed by Nemesis as viewed from Earth. These possibilities can occur four times for each crossing because Nemesis crosses close to each of their orbital paths two times. The palm of the hand is the enlarged, activated and highly-illuminated corona around the star, with the star being the eye. The fingers of the hand are Birkeland column currents reaching outward toward the hidden Jupiter or Saturn.

Thunderbolt and Trident

These plasma displays are highly luminous and energetic. They are plasma discharges of different intensities that occur directly between celestial bodies in space. The *Trident* is arcing probably between a highly-charged and a weakly-charged body. The *Thunderbolt* is arcing sometimes with visibly twisted filaments between two bodies in a close encounter, one of which definitely occurred between Mars and Nemesis. Naturally, as observed from Earth, these events are seen as a profile or side view.

Squatter Man

These very typical and numerous depictions are found in world-wide petroglyphs on cliffs and cave walls and look like a squatting stick man with upraised arms, with or without two donut-like rings on each side of the man's body. The dating of petroglyphs is difficult and their origins may come from Nemesis crossings prior to, as well as after, the Great Flood event of 11,500 years ago. The head of the man is the source of plasma transfer, which is located in the overly charged ionosphere of Earth. The main body stem is the column of Birkeland currents that splits to form a "Y" or the squatter's legs when it comes near a cathode-like planet's surface. The raised arms are unstable sheets of plasma and the two donuts on either side of the body represent a diocotron-type instability. This diocotron configuration is typically seen in laboratories when high energy plasma is emitted between a cathode and anode inside a vacuum tube under certain conditions. The squatter man is the break-down of the atmosphere between the highly charged lowered ionosphere and the very conductive Earth's crust that act like the plates of a capacitor. These archetypes were mostly seen on raised plateaus and mountain ridges closest to the ionosphere. Earthlings are witnessing these events seen in profile, obviously at a safe distance.

Stick Man with Raised Arms

The *Stick Man* is similar to the *Squatter Man* but it lacks the normal donut-like diocotron-type instability configuration. This *Stick Man* has one or multiple raised arms with hands. The small circular hands on the ends of each set of arms are a smaller version of donut-like diocotron-type instability that terminates at the end of sheets of plasma represented by the arms. These plasma displays are likely less energetic than the *Squatter Man* and occur in similar fashion between a lowered and highly charged ionosphere and the ground. These displays could easily be seen by ancient peoples in the light of early dawn or twilight. The continuing hypothesis proposes that mega-corona mass ejections (CME's) were emitted from the Sun due to the closeness of Nemesis. A very powerful CME washed over Earth's magneto-sphere and shrunk it against the ionosphere transferring free ions and electrons. The amount of plasma or electrical charge was so immense that numerous squatter and stick men could be sustained for long spans of time passing through the atmosphere unlike simple lightning that occurs in seconds.

Planet Venus Appearing as a Comet

Now it is time to discuss the very mysterious event of planet Venus becoming a comet. Apparently, interpretations of some ancient texts reveal this episode. Venus, an easily observed wanderer of the ancient skies by the naked eye, appeared without warning as a comet with a typical coma and tail(s). What very possibly occurred is that Nemesis was in the neighborhood and caused an already active Sun to eject an enormous solar flare or corona mass ejection. The plasma was being attracted toward Nemesis when the unfortunate planet of Venus intercepted this high-energy, concentrated plasma, possibly even saving Earth from the same fate. The plasma created a coma or highly energized electrical field around Venus that then struck the Venusian surface with wicked lightening, causing the sputtering and ejection of surface materials. This ejected, dusty material had enough energy to escape Venus's gravity field and appear as tails of a comet as these materials trailed into space behind the planet's orbiting trajectory.

Also, Venus could have produced several tails of ejected materials or received violent arc strikes from an overly active Sun. This vision in the sky could have also created the archetype the *Radiant Crown*.

Celtic Cross and Other Crosses

The *Celtic Cross* is especially the best cross for revealing the Nemesis star. It shows a representation of the *Cosmic Wheel* that is known now to be the feared and revered star, with its spokes and rim. As in all other crosses, the shorter member represents the trajectory of the Nemesis star. The longer member of the cross represents the line of sight from Earth at ninety degrees to Nemesis's crossing through the solar system. This idea really originated with Zecharia Sitchin, author of *The Twelfth Planet*. Apparently, this concept was acquired and re-used by the Christian religion taken from previous "pagan" beliefs to become the cross of Jesus Christ. This religion, as well as others, embellished, at numerous times, the meaning of ancient man's ideas of other important astronomical events such as the spring and winter solstices. These solstices for Christians became Easter (the arisen Christ) and Christmas (the birth of Christ), respectively. Important events in the so-called mythical world are rearranged, renamed, sanitized and utilized to meet the means and ends of the newest culture in power. One must always be careful not to literally interpret ancient epics and cultural paradigms. Man's records of his history have too many times become unrecognizable regurgitations. The original sentiments of record become very mysterious and mythical. Any possible scientific explanation is made very difficult.

Fiery Serpent/Dragon with Long Scales or Feathers

This archetype is typically destructive and feared. The dragon's feathers are dendritic in appearance, which reminds plasma scientists of its electrical nature. And, like a serpent, this archetype is dynamically curving. The curving body is caused by an unstable Birkeland current, trailing a highly-electrified comet or asteroid or sputtered surface material that is rapidly moving through the Earth's atmosphere. This archetype is also represented by multiple dragons or a many-headed dragon traveling through the atmosphere. This visualization is simply the break-up of one larger meteor. These near-Earth asteroids were possibly created by Nemesis and its planets crashing through the Main Belt of asteroids. And, possibly the Dark Planet that had a close encounter with Earth was accompanied by asteroid-like satellites that were stripped off and plummeted to Earth. These falling, burning asteroids with red, dusty trails could have been the fiery dragons. Also, postulated were Z-pinches of Birkland currents snapping or breaking apart as the Cosmic Wheel entourage moved farther away from Earth.

ISSUES WITH THE SUN HAVING A BINARY PARTNER

This dynamical celestial model using a brown dwarf star orbiting around the Sun does have some issues. If this comparatively rapid cycling Nemesis star does frequently create calamity with the Sun's inner planets including Earth, how then did different sets of dinosaur species survive for millions of years without interruption, as the fossil record reveals? This question is answered in one of two ways.

The first way is to claim that Nemesis did cause havoc that is not always recorded until major effects show up in random and infrequent mass extinction events or in Ice Age starts and ends. Life is reputed to get started about 3.8 billion years ago even without a protecting atmosphere. Earth's life forms are hardy enough to regain a foothold and keep evolving one step farther each time after a major catastrophe occurs because the planet presumably remains in the required habitable zone around the Sun and retains its water and atmosphere.

The second way is to claim that the Nemesis system was recently captured within the last few hundred thousand years. Academics have problems with this claim because supposedly any capture mode occurring millions of years after the solar system birth is virtually impossible. Other stars created close to the Sun's birth location in a star burst event are moving apart too fast to ever become gravitationally connected unless it occurs extremely early, such as the first 100,000 years or less. – so, goes the currently accepted theory. This paper's counter-claim is that due to the power law, the smallest stars, such as brown dwarfs, are the most numerous and yet mostly unseen. Interstellar space has so many brown dwarfs that larger star systems will have no difficulty capturing one.

Of course, the main issue is in finding this orbiting, brown dwarf star and/or other closer, independent brown dwarfs. NASA has performed extensive searches and sky surveys and claims the star does not exist. Their efforts were mostly concentrated on searching more than one or two light years away. NASA is dealing with a paradigm that some Nemesis-type star orbits outside the Oort cloud which is another suspicious paradigm. This paper has listed numerous issues with NASA's research about looking for these ubiquitous stars. Dim brown dwarf stars with little proper motion so close to the Sun and located well outside the

Kuiper Belt are very difficult to see in any type of electromagnetic spectrum or tracked by computerization. NASA should keep looking. The seen and recorded alien celestial bodies of the ancient skies speak for themselves.

Why are scientists after more than 30 years of analysis of the data still in clear denial that electrical and magnetic phenomena are clearly big actors in affecting celestial mechanics and the evolution of planetary surfaces? This paper postulates that the scientific community does not want to make any definitive statements to the public because they lack a major perpetrator for these large electrical discharges crossing interplanetary space. Current accepted scientific thought stubbornly refuses to accept "charge separation" occurring in the vacuum of space which planetary arcing requires. Now let's discuss what was recently discovered and analyzed about comets as highlighted by Thornhill in one documentary, "The Electric Comet" which is now part of The Thunderbolt Project.

XXI.

WILL THE REAL COMET STAND UP?

One of the first hypotheses accepted for comets was the "dirty snowball" model developed by Fred Whipple. The comet was mainly composed of frozen ices including water and primordial cosmic dust from the proto-star disk. These volatile materials would boil off creating comas and tails as they approached the warming Sun. Of course, these comets would soon disintegrate in less than a few million years thereby eliminating their presence. Jan Oort then proposed a cloud of comets surrounding the solar system way beyond the farthest planet that could be disturbed by a close passing star. These disturbances over the entire age of the solar system would continue to create long period comets. Shorter period comets were still a mystery until the recently discovered Kuiper Belt of frozen bodies was thought to be their source.

However, these origins for comets became increasingly uncertain as spectroscopic analysis of Comet Halley and Hale-Bopp indicated crystalline silicate structures. The Star Dust Mission of February 1999 to Comet Wild 2 returned some of its materials to Earth which surprised scientists and revealed a possible electrified environment for comets for explaining the materials that were found. One rationalization, yet unproven, was that comet material was somehow transported from the inner regions of a hot forming proto-star to the outer regions of its planetary system.[56]

Some of the more important data from the Star Dust Mission that intercepted two passing comets and returned to Earth are listed[57,58]:

1. Water ice was not a major constituent of the comet.

2. Sulfide compounds were found which cannot form below 50°C to 200°C; these are temperatures much higher than what exists in the outer solar system.

3. Olivine, a product of volcanism, was found in the comet's dust, which is also known to be formed by lightning strikes on Earth. Obviously, if volcanism was involved then the comet is debris from an impact of an inner planet.

4. Cubinite was found which is only formed in liquid water which cannot exist on a small body in the vacuum of space.

5. Pyrrhorite/Sphalerite minerals were found that require extremely high temperatures for their formation which is a dichotomy for comets forming in the frigid cold of the outer solar system.

6. Dust grains were large and crystalline in nature with only trivial amounts of the suspected cosmic dust corroborating that a rocky planetary surface is involved.

The conclusions from this data are completely mystifying to NASA since they failed the test of various popular theories. There is the complete absence of frozen or liquid water despite minerals that require its presence. Water requires atmospheric or other pressure and cannot exist in the vacuum of space. Olivine could not have formed or even survived in the presence of liquid water. Olivine's composition needs extreme selective heating that is only possible on planets in the habitable zone of a fully developed Sun. Dusts with crystalline structure cannot be formed in the frigid vacuum of space. The concept of compositional zoning within the proto-star disk fails the test by this data. The theories of the "dirty snowball" and the Oort cloud are practically eliminated. Another theory of comets and asteroids being the primary source of water and other light volatiles for Earth also needs critical re-consideration.

This conundrum of a comet's composition can only lead to an almost certain concept that comets and possibly most asteroids are the residue of both planetary impacts and other phenomena that is proposed by Thornhill – that of electrical arcing, sputtering, and ejection of huge chunks of surface materials between two highly magnetic and electrified celestial bodies in a close encounter. Thornhill states very emphatically that comets are created by high energy electrical arcing events that blast materials from planetary surfaces. He proposes that possibly Phoebus, one of Mars' moons with similar Martian composition, was blasted away from the planet's surface during one of these arcing events. The startling realization is that something else (not the infrequent disturbance of the Oort cloud) is creating comets of both long and short periods in recent astronomical times (within thousands or less than a million years) which is comparably very short for the overall age of the solar system. Could this something be an unknown orbiting rogue planet or a highly electrically charged Nemesis star that arrives periodically into the Sun's inner solar system to sometimes trigger planetary catastrophes?

The long ignored electrical effects of the Sun's charged particles released in the solar winds are now believed to exchange charges on an approaching comet to create its coma and tails. The close-up videos of comet surfaces reveal similarities with surfaces created in industry by electrical arcing and sputtering. The ejections of dust leaving at supersonic speed from the comet's nucleus wander on the comet's surface and are intermittently discharging just like electrical arcs created here on Earth. The vacuum of space along with the Sun's radiant heat cannot create particles from a comet's now known composition. The videos depicting various comets with their erosion features and varying jets are testament to the comet's electrified environment especially as it nears the Sun and the stronger solar winds of plasma.

Man has only learned recently - the effect of the Sun's solar wind and electrical storms in interplanetary space; the Sun's own helio-magnetosphere discovered by one of the Pioneer missions[59]; the Earth's magnetosphere analyzed by artificial satellites; the connection of auroras with sunspots and solar storms; the harmful effects of solar storms on satellite and space station systems; and the comparative sunspot activity with Earth's climate/weather. Perhaps there is cause to fear much more dramatic electrical events with direct arcing between celestial bodies. If the comets are not proof enough, now the previously referenced Thunderbolt Project by Talbott and Thornhill highlights more startling news about findings from space probe missions to Mars. This brilliant analysis of Martian features further establishes the catastrophic electrical and magnetic effects between celestial bodies.

XXII.

REVISION OF MARTIAN HISTORY DUE TO HIGH ENERGY ELECTRICAL EVENTS

My first personal analysis of the topography and elevation maps released by NASA for the entire Martian surface[60] was the idea of a dominate impact of a very large icy asteroid. This impact created the Hellas crater surrounded by the Tharsis bulge or uplands which was created by a huge outpouring of liquefied mantle and impactor materials from the crater. This event created the dichotomy of the higher volcanic plateau and lower smooth Borealis basin of the northern hemisphere. This smooth basin is thought by NASA to be due to another gigantic impact that created a smooth lava flow similar to the mares found on the Moon. The hypothesis of this paper theorizes that this surface is the original or oldest surface prior to the more familiar type of impact that created the Hellas crater. The volatile materials of the impactor then slowly migrated inside the Martian mantle and around the small core to differentiate and resurface on the opposite hemisphere creating very large volcanoes such as Olympus Mons. Of course, the raised surface received not much later in time the asteroid strikes of the Late Heavy Bombardment (LHB) period. Perhaps the large impact on Mars was just part of the early part of the LHB. This paper also postulated that the asteroid-looking moons of Mars are two of the larger chunks created by the impact that never reached escape velocity and began orbiting the planet.

The Martian surface remained mostly unchanged since Mars lacks - tectonic plates, sufficient internal heat, and the influence of tidal acceleration forces that the Earth receives from a closer Sun and a nearby massive Moon. Geological features appear to indicate erosion by water flow and are currently the basis of NASA's currently accepted idea. However, it is questionable that liquid water ever existed on Mars. Its atmosphere never had a chance to develop due to its freezing temperatures, small mass, and lack of a significant magnetic field that cannot protect the solar wind from stripping atoms away from its ionosphere. NASA's Mars Global Surveyor has detected ionized particles trailing into space behind Mars corroborating this atmospheric loss even presently.[61,62] The atmospheric pressure is and has been too low in the past since its last major impact event. For liquid water to form in large amounts before dissipating as vapor makes it impossible for any significant surface erosion by liquid water and mountain wasting to occur. The only real evidence of surface erosion besides the wind, lava flows and landslides are the seasonal melting of the ice caps of mostly carbon dioxide which do not extend too far from their wintertime perimeters. A very basic conundrum arises since no major amounts of liquid water ever existed on the Martian surface. How do the Martian surface features appear to be caused by water erosion? NASA has been working on this mystery for the last numerous years.

This is where Thornhill and Talbott of the Electric Universe group enter the picture and present a resolution which are global electrical discharge events. Their story and reasons based on analogous laboratory experiments and the observations of lightning, comet tails, and auroras is extremely convincing. These EU authors fill in an amazing missing part of Martian history after the Late Heavy Bombardment period that can include the previously predicted history of a major impact by an icy Pluto-size asteroid. NASA scientists fear to embrace any of their ideas for the creation of Mar's bizarre surface features. The NASA scientists cannot comprehend a source for these massive electrical discharges. The current safe mode is to stick with what is known about water erosion and feverishly search for this missing water that may be sequestered and frozen underground and/or hidden in subterranean cavities. Or possibly any water produced and brought to the surface quickly performed it deeds of erosion and was quickly dispersed into interplanetary space. An amazingly huge amount of water for a considerable length of time is needed to produce the largest canyon in the solar system, Valles Marineris that stretches in a straight line across 1/5 of the planet's circumference. Mysteriously, no riverine delta is revealed at the end of this valley. NASA attempts to help their theory by introducing a huge fault-line hypothesis when plate tectonics is practically non-existent on Mars.

David Talbott along with Wallace Thornhill introduces their hypothesis about an electrified Mars in their book, *Symbols of an Alien Sky* – "Episode 2, The Lightning-Scarred Planet Mars".[63] The basic argument is based around massive global electric discharge, much like lightning on a much larger scale, creating the smaller consistent surface features of Mars. Their proposal includes the creation of Valles Marineris and the highest mountains. This paper still holds to the idea that these mountains are principally created by volcanism and then modified by electrical discharges proposed by Episode 2. The Thunderbolt Project authors emphasize that NASA has made no mention of their ideas, but their illustrated laboratory experiments certainly support their claims.

Here on Earth dendritic or treelike patterns are easily observed for lightning in the sky. Lightning strikes also leave evidence of dendritic patterns on rocks and even human bodies. These forms are the electrical breakdown of channels of current into fractals. George Christoff Lichtenberg first demonstrated the differences of these figures on positive and negative charged surfaces now called Lichtenberg figures. A sputtering high energy electrical arc discharge will create scalloped walls and ridges including chains of varying craters on metallic surfaces. The materials on the surface of Mars were also simulated in the lab and produced these same patterns.[64]

The dendritic patterns fanning outward from the main canyon walls of Valles Marineris are very similar to what is observed in the laboratory. These channels do not appear to have analogous water erosion patterns found on Earth. The claim is that a very large concentrated electrical discharge created the entire almost linear canyon running latitudinally close to its equator. The scalloped ridge walls found on Mars for parts of Valles Marineris, the Victoria impact crater, the calderas of volcanoes and the mountain perimeters of these same volcanoes are claimed to be caused by pinched cylindrical currents just as occurs in the laboratory for electrical plasma machining. Standard geography found on Earth does not conform to these persistent and consistent carved forms on the Martian surface.[64]

Hair-like filaments photographed by the Surveyor probe orbiting Mars are found on the branches of numerous dendritic forms in raised relief above flat surfaces that stretch for hundreds of miles on some Martian plains. These forms can be reproduced in the laboratory with electrical discharges on surfaces that have dust which confirms a spreading of sheet-current on top of some Martian plains. No type of erosion or crystalline chemical process can explain these bizarre forms.[64]

The topography map of Mars shows a raised cratered upland in the southern hemisphere and a very flat and smooth surface like a mare surface on the Moon in the northern hemisphere. Talbott points out that high energy discharge more than likely first struck the higher elevations and then flowed as current toward the lower elevations. The electrical current broke down along the edges into fractals to produce typical geometries that can be produced in the laboratory. These geometries have a general trend or predictable phases that occur in a certain direction going toward the lower elevations which occur in the following order:

1. a network of channels
2. large blocks
3. separate angular islands
4. various pyramidal shapes and sharp edges
5. isolated blocks and mounds
6. flat depressions occurring at previously eroded high points.

These geometries are also observed with no surprise on the perimeters of the Martian uplands.[64] The proof of global electrical discharge on Mars keeps mounting.

The Rover Opportunity discovered trillions of objects labeled as blueberries which are bee size spheroids. These little beads are claimed by NASA to be hematite concretion that supposedly can only be produced in the presence of water that is still missing.[65] However, these same spheroidal shapes can be reproduced in the lab by blasting hematite-type soil with an electric arc. Other much larger rounded shapes or domes inside similar sized craters range in size from 100 meters to one mile in diameter. The lab can also produce these shapes, but of course, in a much smaller, scalable size. Dome craters many miles wide are found near the poles on Mars. Only the presence of immense, high energy, columnar-type electrical discharge can cause these features on Mars. Some close encounter of a celestial body with this type of energy has to be the source. And, it is perfectly reasonable to suggest the crossing of a brown dwarf star through the inner solar system. Mars which can sometimes be at the conjunction between the Sun and this passing brown dwarf can easily become the victim of a connecting circuit of solar winds and huge Birkeland-type column(s) of electrical discharge between the two stars. Mars was in the fateful path of lightning bolts being exchanged between these two stars. The other inner planets had periodically similar fates. Earth's encounters with this exchange of thunderbolts received catastrophic effects but definitely not as severe as with Mars due to its farther separation from the Nemesis star.

Talbott and Thornhill also discuss the effects of other major electrical discharge events on other planets and satellites of the solar system, but never really address any possible serious effects on Earth caused by celestial events that were observed, recorded, and compiled in their documentaries, *Symbols in an Alien Sky*. However, this paper does accuse the Nemesis star for catastrophic events occurring periodically here on Earth. The most serious of known recent events occurred during the end of the Younger Dryas Period around 11,500 years ago marking the extinction of numerous fauna and the melting of the North American Ice Cap. When Earth is caught in the conjunction between the Sun and Nemesis, both increased electrical and tidal acceleration events can occur. The side effects are increased volcanism, earthquakes, tsunamis, severe weather, dramatic climatic changes, possible magnetic reversals, and a possible temporary wobble of the Earth and/or shifting of the crust and mantle on the core bringing adjustments to the overall geoid which caused global flooding. Asteroid or comet strikes are other possible culprits. The thought does certainly occur that mankind had previous civilizations that were wiped out in a global fashion one or more times. Then, mankind driven into Stone Age conditions slowly resurrects his developed cultures each time.

XXIII.

ANCIENT SUN-GOD DESCRIPTIONS GIVE PROOF FOR AN ORBITING NEMESIS

The Electric Universe (EU) group of independent scientists have a current champion in promoting comparative mythology along with one of its main founders, David Talbott. His name is Ev Cochrane. I have met Cochrane at an EU Conference. He is very excited about promoting Talbott's ancient archetypes through his own research. Of course, his support includes recognizing Talbott's book, *The Saturn Myth,* and its Saturn polar configuration concept. This book published in 1980 proposes that ancient myths and traditions describe the planet Saturn as a dominant celestial body in the sky and how Earth, Mars, and Venus were once part of the Saturn system. I flatly refuse to accept Talbott's hypothesis and give my reasons in my paper, "Problems with the Saturn Myth's Polar Configuration" (ettingerjournals.com/dbe_saturn_myth.shtml). Otherwise, I enthusiastically support EU's other ideas and goals including Talbott's archetypes of symbols presented in an alien sky.

I accidently discovered one of Ev Cochrane's articles, "Anomalies in Ancient Descriptions of the Sun-God" published in the Chronology & Catastrophism Review 2016:2 of the Society for Interdisciplinary Studies (SIS). He wrote superb, delineated, and understandable descriptions of how the Sun-Gods of ancient Mesopotamia, Egypt, and India differ with our Sun today. Of course, I was especially interested in his threads connecting the polar configuration using the science of celestial mechanics and astrophysics. He did not make any such connection. His adequate conclusion was that the Mesopotamian and Egyptian testimony with regard to physical description and accompanying features of the ancient Sun-God is taken quite literally as astronomical reality. I totally agree with Ev Cochrane. This perspective and conclusion are totally different from any scientific orthodoxy which only attempts to explain these anomalies metaphorically or mythologically. However, this paper will explain Cochrane's so-called anomalies as events that really did happen in the sky based on the Nemesis star that returns approximately every 3600 years. Cochrane's descriptions or translations are used as evidence that a second star, the Sun's brown dwarf sister star does make its presence known periodically to mankind in the manner that is carefully portrayed by Cochrane.

Unlike Velikovsky and Talbott, Sitchin translated and interpreted the ancient languages directly from clay tablets and stone scrolls found in the ancient libraries of the Sumerians, Babylonians and Akkadians that are now dispersed to modern-day museums throughout Europe and the Middle East. These texts included much of the sources cited by Ev Cochrane: "The Epic Tale of Creation by the Sumerians"; the Gilgamish Epic; the Babylonian creation text known as the "Enuma Elish"; Sumerian hymns celebrating the Sun-God; the Akkadian cylinder seals which depicted the Sun-God; the Pyramid and Coffin Texts of Egypt; and the early Egyptian Dynastic Shamash hymns. Sitchin had the unusual talent of directly translating the Sumerian cuneiform, and its later adaptations of Egyptian hieroglyphics and Hebrew languages. Sitchin did not use other people's interpretations who incorporated many ill pre-conceived notions and lacked the necessary language skills.

Sitchin's storyline is truly amazing. A people called the Anunnaki came from another planet that periodically orbits the Sun. These Earth aliens settled on this planet 200,000 to 400,000 years ago. These astronauts created us from the Earth's existing primates and their own DNA. Their stories are what is handed down to the Sumerians in such Epics as the Tale of Creation and Gilgamesh thousands of years before the dating of Sumerian civilization; they taught their ideas sometimes metaphorically for easier understanding to humans about a second Sun, their home planet called Nibiru, the Sar cycle, and other astronomical concepts. When I first read Sitchin I thought his translation was sheer nonsense. At first reading I was only interested in his translated reason for the Great Flood. The well-known Noah story where the Gods warned him about a pending flood is an abridged story that the Hebrews took from the Babylonians who received it from the Sumerians and then finally from the Anunnaki. Sitchin was revealing the real source of the Biblical stories. Then the gaps in understanding began to close when I further educated myself about the solar system mysteries, the Moon enigma, the Electric Universe concepts, brown/red dwarf stars, and plasma archetypes identified by David Talbott and Anthony Peratt. The perception by ancient humans were that the Anunnaki were gods, and the planets and the stars were Anunnaki's Gods who fought battles in the great sky arena.

The appearance of battles was due to electric discharges in the plasma glow mode being exchanged between the planets of both the Sun and Nemesis, and could be witnessed most of the time from a safe distance by the inhabitants of Earth. When Ev Cochrane refers to a special, strange object that rises and sets in the sky as a Sun-God, he unknowingly is talking about Nemesis and its archetypes or symbols described by the ancients. What caught my eye when first glancing through Cochrane's article was a figure (his Figure 5) depicting typical cylinder seals that first appeared in the fourth millennium BC. I immediately recognized the depiction as the phases of Nemesis when it passes through the inner solar system above the ecliptic plane displaying all its different phases from Earth. A picture of these cylinder seals with various crescent orientations taken from Cochrane's article follows. A sketch then follows showing how these crescents or phases occur for Nemesis as it crosses through the solar system.

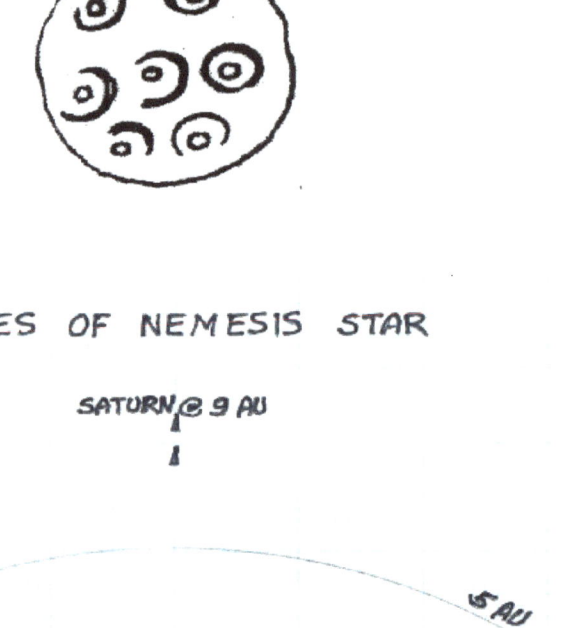

Figure 26: Diagram Depicting Phases of Nemesis' Corona When Crossing Through Inner Solar System

The above sketch roughly shows to scale the probable orbital path of the brown dwarf star, Nemesis, with a periapsis of 3.0 AU crossing the solar system between Mars and Jupiter above the ecliptic plane. Earth is shown at different locations during its yearly orbit. Depending where Earth is in its orbit affects how the different phases of Nemesis are seen.

Cochrane in his article delineated ten reasons why the ancient Sun-God was not our familiar Sun. He is absolutely correct - these sky symbols were Nemesis, the home of the Annunaki who wrote and sang hymns of this homecoming event. There are always two different Sun-Gods when Nemesis makes its crossing through the Sun's planetary system. The Nemesis Sun-God, their home star, is most important to these ancient astronauts. So, now let's examine more carefully Cochrane's ten reasons by using many of the EU's scientific premises and the Sitchin storyline.

His first reason is about ancient accounts that the Sun-God's sunrise is routinely portrayed as accompanied by the shaking of heaven and Earth. Immediately, I must digress to explain what the sunrise really means. Nemesis is arriving into the Sun's planetary system by first crossing the Sun's helio-sheath and coming into view by Earth inhabitants after 3600 years. This entrance is a very special Sunrise. Then the special Sunset is the departure of Nemesis through the helio-sheath on the opposing side of the solar system. Of course, there are the normal sunrises and sunsets due to the rotation of Earth every day. Nemesis and the Sun may rise coincidentally for many days and then their sunrises will begin to occur differently as the brown dwarf star moves through the solar system getting closer to the orbital paths of the planets of Mars and Jupiter.

In EU theory a typical brown dwarf has its own electrical circuitry with the galaxy, just as the Sun does. It may on many occasions collect more charge than the Sun for a particular orbit. When it intersects the helio-sheath, a double layer of current connected to the Sun's polar regions will drain electrical energy from the Nemesis system onto the heliosphere's double layer and then abnormally energize the Sun. Next, the Sun produces increasingly more solar wind by emitting more protons, electrons, and other ionic particles. When the Earth in turn receives this super-elevated solar wind, the ionosphere is first affected and then Earth's crustal circuitry is also energized. The affect for humans is increased lightning due to more capacitance between ground and the clouds and increased crustal activity such as earthquakes and volcanism due to more capacitance between the charged Moho layer and the conductive waters on the crust's surface. Hence, the ancients associated the Sun-God's coming or its first Sunrise with the 'shaking of heaven and Earth'.

This first Sunrise of the Sun-God is also associated with the 'occasion of great tumult and noise'. Because the solar winds increase and are more charged, the Earth's weather is affected as solar meteorologists are currently learning. This heightened weather will bring excitable auroras, severe windstorms, tornadoes, and thunderstorms that may occur especially as the Nemesis's charged corona completes its passing through the helio-sheath. These tumultuous weather conditions then subside and disappear as Nemesis completely passes through the helio-sheath region and moves more toward the center of the heliosphere bubble. Once again, the helio-sheath is crossed on Nemesis's departure. Most likely, there is no charge transfer during Nemesis's exit since equilibrium of charge energy was already achieved after Nemesis crossed the helio-sheath the first time.

The ancient technocrats, with much more prowess in astronomy than current historians could ever expect, knew Nemesis would return. This momentous appearance would be both foreboding, due to possible planetary battles, and rejoicing, for their ancestry were returning and they could have a much-anticipated reunion with their home planet and possibly with its inhabitants every 3600 years. Their knowledge of astronomy would lead them to use markers, either natural, such as mountains or hills, or erect their own markers, such as obelisks or pyramids, to show first sighting of the returning sister star. First appearance could have been marked with the conjunction of two sunrises, *both Nemesis and the Sun coming over 'twin-peaked' mountains* or between some man-made markers.

Naturally, Nemesis will move across the sky just as planets do today over several years of time after passing through the semi-latus rectum of its elliptical path with the Sun. I estimate that the Nemesis system takes several 10's of years to cross the solar system between the two semi-latus rectum points where its phases can be seen clearly by mankind. The time span for Nemesis to pass through the Sun's heliosphere is estimated at 250 years by knowing the orbital periods of the outer planets. Its path across the sky will definitely be different from the Sun's path, especially since it is considered to have a steep inclined orbital plane from the ecliptic plane of the Sun's planets. Nemesis may have been large enough, given the diameter of its corona envelope, and bright enough to be seen in the daylight sky, just as our Moon today. Hence, after sunrise of the Sun occurred for certain orbital positions, Nemesis would *make a daily appearance from the 'heart' or middle of the sky*. This sudden appearance in the sky could also happen in the night sky. These appearances more than likely occurred between the semi-latus rectum points; beyond these points, Nemesis may be too far away to see as more than a large point in the sky.

One of Cochrane's anomalous but very important reasons is erroneously connected to the polar configuration of the Saturn myth. The 'Saturnists', who he accepts and promotes, suppose the *'horned' celestial body* is due to a reflection of Sun's light off the limb of Saturn forming a crescent which seen from Earth is in a polar configuration with Saturn, Venus, and Mars. Yes, the crescent is a reflection due to the Sun's light reflecting off the limb of the corona envelope of the brown dwarf star, Nemesis, but not Saturn. This corona features a distinct boundary similar to red giant stars in the galaxy, but is much dimmer. The light from the star's core is bright enough to shine through the boundary for observers to see who would be close enough, such as people on Earth during Nemesis's crossing. The crescent shown on the corona perimeter appears to the less educated humans as the horns of a bull or possibly the wings of a bird. The orientation of the horns changes as the configuration of the Sun/Earth/Nemesis system changes just as the phases of the Moon do for earthlings.

Cochrane points out that Akkadian cylinder seals depict the *Sun-God appearing between two 'gates' or 'doors'* and sometimes the 'twin peaks' are placed between these same gates. These gates are also associated with the *double or back-to-back bull in the sky* which is considered to convey important reality about the visible sky. This double bull presentation appears in predynastic artworks dating to 3100 BC which demonstrates this astronomical concept comes from much older times than what is considered the peak of the dynastic Egyptian civilization. Obviously, the ancient technocrats initiated this civilization and probably had already built the Giza pyramids. So how are the gates or doors associated with the back-to-back bulls? This presentation is no metaphor; it simply represents the whole storyline of Nemesis crossing through the solar system. The gates represent Nemesis passing through the heliospheres of the Sun creating a plasma glow display that could be seen faintly from Earth. As Cochrane points out, the Coffin Texts of Egypt link the double bulls to 'brightness' and an early Dynastic Shamash hymn declare that the bisons of Shamash make a divine radiance. This 'brightness' is the plasma glow displays that occur at the 'gates or doors' as Nemesis enters and exits the solar system through the helio-sheath of the Sun. Nemesis's approach from one direction would eventually reflect the Sun's light as a crescent pointing in one direction away from Earth. As Nemesis journeyed to the other end of the solar system the crescent would be pointing in the opposite direction away from Earth. These opposing crescents are exactly what was being embodied by the horns of the back-to-back bulls. And, the opposing crescents or horns occurred near the gates or doors after entering and before exiting because of the direction of the Sun's rays reflecting off the limbs of the activated corona.

The ancient Sun-God is described as being *'fixed in the midst of the sky'* and unmoving. Of course, Nemesis is moving through the solar system but appears to be very slow compared with the background of the stars and the Sun's planets. Nemesis is moving in a retrograde orbit in the opposite direction of the normal planets and at the region of its periapsis the orbit appears almost stationary from Earth's view likened to watching the complete trajectory of a ball thrown very high overhead. Nemesis is never seen in its complete orbit, but only a very small portion that occurs within the Sun's heliosphere. Also, Nemesis crosses the helio-sheath about 100 AU from the Sun and will appear very small and insignificant and fairly stationary due to an extremely elongated elliptical orbit. Its size and brightness will increase enough to see the details of its phases when it roughly reaches the orbital distance of Jupiter. Hence, the brown dwarf appears almost unmoving for the many years while it is viewed inside the heliosphere.

In ancient tradition the locus of sunrise and the locus of sunset are identified to occur together. This interpretation can only mean that the loci of the sunrises and sunsets of Nemesis and the Sun occur together when Nemesis is at roughly a right angle to the line of sight between the Earth and Sun. At that time both stars would appear to have a similar locus of both sunrise and sunset. Similarly, when Nemesis is exiting the solar system it will produce a right angle to the line of sight between the Earth and Sun again. Then, as before the two stars viewed from Earth will have a similar locus for both their sunrises and sunsets. This situation is similar in today's solar system with the viewing of the well-known morning and evening stars, Mercury and Venus, at sunrise and sunset when their orbital positions are configured properly.

Arriving at Cochrane's Number 2 reason are ancient accounts of a typical sunrise being accompanied by a 'greening' of the skies. A traditional Egyptian passage is cited, 'Hail to you who rises in turquoise.' Cochrane provides no answer for this phenomenon except to say that this does not occur today with the current Sun. I will provide an answer from the viewpoint that a brown dwarf star enters the solar system every 3600 years through the doors of the helio-sheath.

What happens is that the intersection of the two stars' magneto-sheaths activates increased electrical current flow that energizes one or both of the stars that then release increased amounts of charged ions and electrons into the ecliptic plane that then impinge on Earth and other planets. This ionosphere of the Earth receives much more than the average charged particles from the solar winds. These excessive amounts of charged particles not only lower the latitude where auroras are normally seen but also ionizes and excites the higher atmospheric constituents throughout most latitudes. Then emissions of light of varying color and complexity are created. 'Aurora' comes from the Latin word for 'dawn, morning light' since auroras were formerly thought to be the first light of dawn. This is the precise time when the ancients saw a 'greening' of the cosmos.

The oxygen element at the top of the atmosphere is excited and emits at 630 nm in the color of red. The low amount of oxygen atoms produces a faint appearance of scarlet, crimson, and carmine that are hues of red. However, at lower altitudes the more frequent collisions suppress the 630-nm mode and the 557.7 nm emission of green dominates. The fairly high concentration of atomic oxygen and the higher eye sensitivity in green make green auroras the most common. The excited molecular nitrogen found at this same altitude transfers energy by collision to an oxygen atom that then enhances the emission even more in the green wavelength. The prediction made is that extended green auroras in a more diffuse and steady appearance occurred at much lower latitudes for Egyptians and Mesopotamians to see due to the arrival of Nemesis crossing the Sun's magneto-sheath. This 'greening' was probably not seen during Nemesis's sunset or departure from the solar system since most of the energy transfer between the two stars had already taken place during the initial contact of their electrified magneto-sheaths. The following typical diffuse aurora that is dominated by green is created by a greater solar wind and increased ionospheric activity when Nemesis passes through.

Figure 27: Depiction of a Green Aurora

Cochrane's final, and to him the most important, reason is that archaic pictures engraved on cylinder seals and stone reliefs do not represent the current Sun's orb. Some of these most plentiful and well-preserved representations show a circle with a dot in the middle, a crescent with a dot in the middle, crescents oriented in different directions, four-pointed or eight-pointed stars inside a disc, and various combinations of all these elements. Cochrane further explains that the very intelligent Mesopotamian civilization that invented writing, the calendar, numbering systems, and scientific astronomy could not be possibly wrong in their depictions. Ev Cochrane is absolutely correct. The 'twist' for me is that these people were much more intelligent than historians could ever imagine. These ancient people are possibly ancestors of alien astronauts that came from a planet called Nibiru (from the Sumerians) that orbits the Nemesis brown dwarf star. This unbelievable storyline comes from Zecharia Sitchin and his translation of ancient cuneiform and other languages. I do not expect many readers to consider this as possible. So, it is OK for any readership to disregard this part which I call my 'twist.' However, please pay attention to my explanations of the listed anomalous descriptions of Ev Cochrane's Sun-God which I claim are neither the Sun nor the proto-planet Saturn of EU's polar configuration fame. Which model do you think makes more sense from a celestial mechanics perspective? *Planets, including Earth, aligned in a polar configuration with Saturn's pole OR a brown dwarf star orbiting the Sun every 3600 years with its periapsis located somewhere between Mars and Jupiter well above the ecliptic plane. Admittedly, both models are perhaps beyond the scope of most intelligent thinkers who cherish their mis-leading and muddled paradigms. But you the unbiased reader/scientist are now given the challenge of deciding on the best model for a proven, mysterious, ancient sky.*

The subsequent explanations are given for the Sun-God symbols and/or Talbott's archetypes in the following diagram.

1. The circle with a dot is the view of a brown dwarf star with its corona representing the circle; and, its inner spherical plasma core is shining through its outer perimeter of the dim plasma to manifest the central dot.

2. The crescents are the reflection of the Sun's light being cast off the roughly spherical envelope of the Nemesis star, which includes the solar winds interacting electrically with its perimeter.

3. The orientation of the crescents are the different phases of Nemesis, similar to the Moon's phases. The phase that is shown on the bottom side is when the brown dwarf on its highly inclined orbit is well above the ecliptic plane near its periapsis. The back-to-back bulls represent Nemesis' phases during both its arrival to and departure from the solar system.

4. The 4- or 8-pointed star inside the disk of the Nemesis star disk is either an occulted planet of Mars or one of Nemesis's own planets. The occasional wavy arms of this inner star may represent the brown dwarf star flaring as modern astronomers know that happens to some recently observed brown and red dwarf stars.

5. One particular premise of the Electric Universe (EU), which I support, is that dim brown dwarf stars generate enough heat and light to support life in a liquid-water zone by providing a protective corona of plasma almost in the dark mode. EU's idea of a Golden Age is the time span that Earth was inside a protective cocoon or corona of Saturn when it was still a brown dwarf star captured by the Sun - whereas the Golden Age from my perspective is the time Earth's climates were

friendlier, before the Great Flood of 11,500 years BP, when the Annunaki were in charge of man's civilizations and their building of megalithic structures. The flood event destroyed most of their infrastructure thereby allowing their human creations, mankind, to take on more responsibility. Another possible interpretation of a Golden Age would be the time the alien astronauts, the Annunaki, spent on their home planet with its constant daylight and mild unchanging climate.

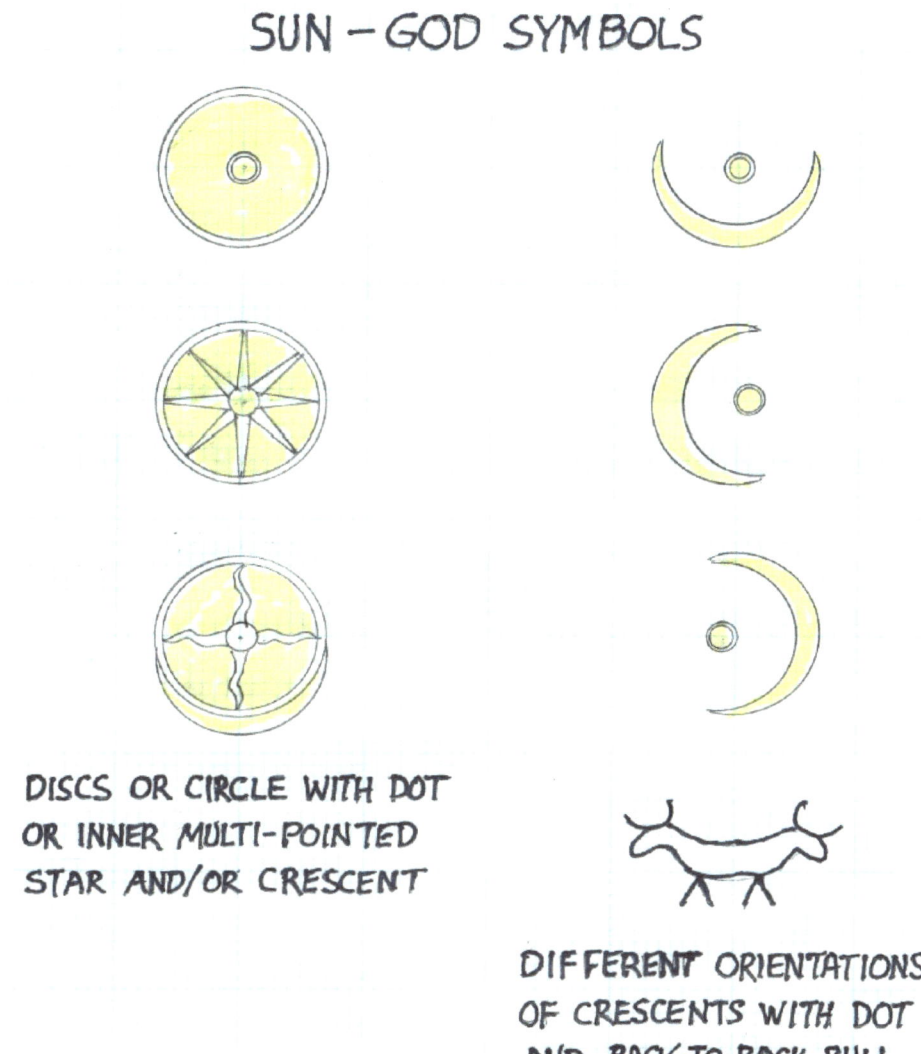

Figure 28: Symbols of Nemesis as the Sun

These symbols or depictions are prolific and typical throughout ancient (6000 to 2000 BC) Mesopotamian and Egyptian artwork found on clay tablets, stone scrolls, cylinder seals, rock reliefs, and rock sculptures.

Ev Cochrane is intent in over-turning the orthodox view of the present solar system. In my estimation, he is unable to formulate a model that is acceptable by the standards of celestial mechanics which modern astrophysicists know works. Celestial mechanics based on gravity and kinetics is successfully applied to space travel. Astronomers do know from recent observations and discoveries that brown dwarfs do exist, and are very plentiful, and have planets of their own, and have extinction sheaths that cause dimming and reduced surface temperatures, possibly undetectable by the best infrared telescopes, and orbit other larger stars. All outer-planet satellites were either created in a binary fashion by their parents 'spitting out' moons from their equatorial regions; or, these moons were accreted from dusty disks in their parent's equatorial plane. No model of planets having a polar train has ever been discovered or conceived by consensus science – thus discounting the Saturn polar configuration hypothesis.

Astronomers do **not** know of any existing planets or even stars with polar configurations and how they could possibly be assembled in that configuration and then re-arrange themselves into a normal star system as orbiting planets. But the EU establishment keeps insisting that Saturn's polar configuration did occur; my hopes are high that this very enlightened group will reconsider their own paradigm. I wish to thank Ev Cochrane for his careful research into anomalous Sun-God descriptions. He certainly has corroborated my model of Nemesis, the orbiting brown dwarf star. I pray that we both remain open-minded and that we may have enjoyable discussions together in the future about our different concepts. Perhaps, I can convince Ev that he needs a better model for his masterful findings.

XXIV.
THE SCIENCE OF PETROGLYPHS
BY ANTHONY PERATT

The science of petroglyphs as studied by Anthony Peratt provides conclusive evidence that Earth has been confronted and challenged by intense solar outbursts in antiquity and in prehistory. This evidence supports the connection between the coming of Nemesis and the final step of a global flood to complete the catastrophic storyline occurring 9500 years BC. The data that Peratt has compiled and written about is found in Publications of the IEEE Nuclear and Plasma Sciences Society beginning in 2003.

These worldwide petroglyphs in their many varied forms depict some of the following: funnels of concentric sheets of Birkland currents; conical inflow of current with striations on the body; helical instabilities sometimes shown as caterpillars; flattened toroid (coils shaped like a donut) seen in both front and oblique views; eye-masks or owl eyes which are the cross-sections of a toroid found at the terminuses of Birkland currents; merging crisscrosses and separatrix patterns; squatter man without and with diocotron instabilities depicted as symmetrical dots or small circles on each side of the body; horn-sheep and mushroom profiles; Stonehenge reconstructions with circular dots or filaments of 56, 28, 6,5, and 4 (converging to 4 as current increases in laboratory experiments); and, the Windjana eclipses (Australian name) with concentric circles and joining parallel lines. These pictures or archetypes are found at numerous and distant locations including different continents. They are amazingly consistent.

All these ancient symbols drawn by humans are associated with high-current Z-pinches produced in the laboratory experiments. The fantastic similarity of the petroglyphs and geoglyphs with plasma experimental morphologies is excellent proof that these same plasma displays were reproduced in the ancient skies and recorded by mankind. The details of these experiments at the Los Alamos National Laboratory and the cataloging of worldwide petroglyphs on each continent can be found at www.plasmauniverse.info. The petroglyph descriptions, locations, line of site, and possible dating are listed.

Peratt agrees with T. Gold's ("Large Solar Outburst in the Past,"1962, Pontificiae Academiae Scientiarvm Scripta Varia. Vol. 25, pp 159-174) hypothesis that a catastrophic event of a solar coronal mass ejection of 1 to 2 orders of magnitude larger than those observed today reached Earth and created an intense Z-pinch around the planet. This Z-pinch allowed currents flowing in the same direction as 56 filaments turning into 28 and eventually into 4 per Biot-Savart force laws for currents flowing in the same direction. In turn, relativistic electrons were directed toward the south pole creating an aurora looking like the petroglyphs of the Windjana eclipses of Australia. Hypervelocity protons were directed toward the ancient axial north pole creating aurorae that looked like the 28-ray/nose-and-eyebrow-facemask petroglyphs of the United States Northwest. A strong circular aurora of today may have a total current of 7 mega-amperes with 28 filaments containing 250 kilo-amperes each. Gold estimates a thousand-fold increase or more occurred in antiquity. Each filament would have carried 250 mega-amperes with a total current of 7 giga-amperes. The possibility of even more than another order of magnitude of electrical current is discussed later which jerked the dipole magnet of the Earth's crustal/mantle shell (CMS).

Interestingly, Peratt surmises that the very long straight Nazca lines of Peru represent glowing filaments running longitudinally across the sky. I like Peratt's and Gold's ideas, but they fail to link this electrical catastrophe with any crustal/mantle shell displacement or the global flood. These scientists also fail to connect any significance to the Cosmic Wheel archetype recorded in antiquity and to the possible re-occurrence of this type of catastrophe in lesser degree for each of the Nemesis-star's periodic visitations to the inner solar system. These plasma displays in Earth's skies may have occurred many times but were easily forgotten due to the long span of time between each crossing of Nemesis. However, the real memories are brought to light by Peratt's study of petroglyphs and laboratory plasma experiments. These memories are further enhanced by David Talbott's complete study of archetypes in his "Symbols of an Alien Sky" previously discussed. Postulation of other visits by Nemesis since the Great Deluge is certainly attested by Talbott's collection of archetypes which encompass not only the times of petroglyphs but also the ancient times when art and language of numerous failed civilizations revealed electromagnetic catastrophes occurring after the Great Deluge.

Peratt has suggested that the recording by witnesses of these intense solar outbursts lasted many years due to the geographical range and consistency of the drawn petroglyphs. This rock art is truly authentic and has not been defaced or altered through several millennia. Why did these plasma displays last so long to be recorded easily by men's memories? The hypothesis is that a mega-coronal mass ejection (CME) having an energy equivalent to 1×10^{29} ergs - even larger than what Gold proposed washed over the magnetosphere of the

Earth and shrunk it next to the ionosphere. The magnetic field strength was so strong that it quickly jerked and aligned the already magnetized, tilted crustal/mantle shell (CMS) about the inner core. These magnetic vectors caused the CMS to rotate 25 to 30 degrees longitudinally thereby changing the geoid shape of the crust. This event, in turn, caused global topographical changes, moved the location of the axial poles on the CMS, and sank Antarctica's land mass causing its ice sheet to slide into the ocean and create a global flood.

Surprisingly to Peratt, is that many rock art images are only found in extreme energy density experiments; no other morphologies for lower densities are observed. Any column of plasma showing both sausage and helix patterns from intense currents can only originate from the Sun if the solar flux increased one or two magnitudes. Or, another source of plasma entered the solar system. This discussion brings forward the strong probability if not actuality of a brown-dwarf Nemesis star. This brown dwarf is postulated to both energize the Sun to emit abnormal solar winds and mega-CME's; and, also emit via planetary conjunctions an immense plasma discharge on the axial north pole of Earth. The dating of the previously discussed rock art was determined to be in the range of 10,000 BC to 2000 BC which corroborates the dating of the Great Deluge and other later visits by Nemesis. Peratt claims that buried horizontal petroglyphs in New Mexico and Australia were beneath carbon from campfires making radiocarbon dating possible.

Peratt does not fully comprehend the enormity of this event and how the ionosphere became super- charged like a capacitor plate which bled off charge randomly and frequently over a large span of time to the Earth's surface. These plasma discharges would occur much more frequently along mountain ridges and high plateaus and tectonic boundaries that acted like lightning rods that also provided more conductivity than surrounding regions. Yes, the ionosphere in its very active plasma state would take perhaps several years to reach today's equilibrium. Birkland currents in the form of stick man removed the excess charge and magnetic field strength either by discharging to the Earth's crust or discharging back to the restored magnetosphere as the ionosphere began to rise.

The many petroglyphs portraying stick man and squatter man are indeed the random plasma discharges between the ionosphere and the Earth's crust similar to cloud-to-ground lightning of today. Peratt tries to incorrectly connect all the petroglyph sights to the observed aurora regions of either the axial north or south poles not realizing that stick-man archetypes are local plasma discharge events which are sputtering away crustal surfaces to form unusual landscapes. These stick men created the mysterious landscapes of the United States Southwest such as mesas, buttes, pinnacles, hoodoos, and arches. The energy of the plasma discharges striking the ground excavated even layers of rocks and deposited them into other even layers to recrystallize again. The electrical energy heated the very conductive subterranean aquifers and molten rock in tectonic boundaries to produce gurgling hoodoos and bubbling holes or arches above the surface. None of these strange topographical characteristics could have been produced by wind or water erosion as is still believed by many scientists. The stick men of antiquity moved across the land appearing as alien, fearful god-like beings destroying everything under their feet. These sky-high beings of unknown origins would be dutifully and reverently recorded as rock art by the few lucky human survivors.

Please review a complete charting of petroglyphs discovered and compiled by the astounding efforts of Peratt's team of scientists. Each hemisphere shows the locations of these pictorials. Local maps indicate that most of these sights are in mountainous and elevated plateau regions or volcanic islands as would be expected. The reference for this data is the IEEE Transactions on Plasma Science, Vol. 35, No. 4, August 2007.

XXV.

DOES HUMANITY REALLY HAVE LOST CIVILIZATIONS AND/OR ANCIENT ASTRONAUTS?

One more author will be referenced before summarizing all the facts and ideas presented. This author is Erich von Daniken who wrote the infamous and best-selling *Chariots of the Gods* first published in 1968.[66] Von Daniken was chiefly responsible for popularizing the ancient astronaut hypothesis that was rejected categorically by scientists and academics as pseudo history and pseudo archaeology. He was accused of many factual and logical errors throughout all his books. Nevertheless, he popularized in public thought the mysteries of why ancient megaliths and pyramid-type structures were built. And, also importantly, he asks the question of how were these structures built? The public became more aware and questioned the Egyptian pyramids, Stonehenge, the Moai statues of Easter Island, the Nazca lines of Peru, the appearance of machined stonework in Bolivia and close-fitting stones of Inca walls.

Von Daniken made claims that human beings of ancient times did not have the infrastructure, tools, and knowledge to build such structures. These structures were made either wholly or by guidance from the superior knowledge of aliens of other worlds. An interesting attack was made by John Flenley and Paul Bahn who suggested that Von Daniken's interpretation of the Easter Island statues "ignores the real achievements of our ancestors and constitutes the ultimate in racism: (he) belittles the abilities and ingenuity of the human species as a whole."[67] There existed a very strong prejudice against alien astronauts especially coming from religious elements. This deep-seated prejudice has been greatly reduced since those times due to mankind's own recent interplanetary space travel. This emotional criticism still does not account for all these amazing stone monuments found worldwide that are supposedly three to four thousand years old and perhaps much older.

One very plausible idea that Von Daniken fails to suggest is probably because it is less dramatic in his eyes and will not sell books as well. Perhaps there were no ancient astronauts, but mankind's own advanced civilization(s) that were destroyed due to catastrophic events that only left the most durable texts and largest structures or monoliths behind. Scientists today understand very well that asteroid strikes on Earth can trigger flooding, volcanism, earthquakes, severe climatic changes and electrical storms that can wipe out man's present civilizations. There is good reason to think that global catastrophe happened one or more times in the past to reduce mankind's achievements to a few monoliths and destroy his collective knowledge with only memories preserved mostly by verbalization now identified as myths. The survivors had to re-emerge and evolve their knowledge once again while trying to formulate reasons for ancient texts and structures that are found on every continent. We may be the latest human survivors believing that our myths are just constructions of the imagination, similar to our current religions, and not real events.

Many of Von Daniken's claims of ancient artifacts are very believable, and respectable academics and scientists have been researching at a very animated pitch to debunk his claims. Some of this so-called successful debunking has only occurred in the past 10 years with the best designers, structural engineers, and computerized CAD/CAM systems. What took so long? The task was extremely difficult and many of the proposed tools and methodology have not been tried completely in the field or laboratory. Numerous proposals have already themselves been debunked. Why should the ancients with their limited technology and infrastructure even chose to quarry, cut, transport, and lift into place such huge stones some being as heavy as 60 to 100 tons? Modern day construction including that of the Romans chose more reasonable sizes and weights for their largest pieces of structures (excepting the Baalbek structures where supposedly certain mid-size monoliths were moved in a limited fashion by Roman cranes[68]). Choosing such megalithic stones would be considered rather stupid today unless perhaps there is a greater reason than mere function. Maybe the builders had in mind the maximum durability and lifetime of hundreds of generations considering that Earth's surface has a cyclic hostile environment. Maybe structures such as the pyramids served as some unknown type of machine. Let's review in some detail some of the best-known claims by Von Daniken.

XXVI.

THE POSSIBILITY OF ANCIENT FLYING MACHINES

One claim is that in ancient times a knowledge of making and using flying machines was prevalent and then lost. Ancient texts refer numerous times to machines arriving from skyward. One particular reference is the Vimana or flying palaces mentioned in Hindu mythology. This self-moving aerial car or throne moved occupants through the air and was originally made by Vishwakarma for Brahma, the Hindu god of creation. These vehicles were distinct from predecessors of flying horse-drawn chariots mentioned in the older Sanskrit epics.[69] Numerous stone friezes and temple shapes depict these flying machines. The Sanskrit translations are quite literal in describing devices carrying occupants across the sky. Can this consistent line of thought really be attributed to the imagination or are these Vimanas embellishments of long lost existing flying machines?

The Nazca Lines, a series of ancient geoglyphs, located in a high, arid plateau of Peru, are considered to require ancient aerial feats in order to be produced and observed. The lines are shallow and made in the ground by scraping red pebbles away to expose white limestone that eventually hardens and accents the designs for aerial views due to their size of 400 to 600 feet in width. The designs depict more than seventy animals such as birds, fish, lizards, monkeys, and human figures.[70] Von Daniken's claim is that only aerial devices could aid in their making and their viewing. Some debunking proved that these geometric and artistic designs could easily be planned and made on the ground without aerial aids although someone has conjectured the use of balloon flight. Also, some but not all the designs can be viewed from nearby hillsides. The best viewing is definitely from an airplane. Archaeologists believe that the Nazca culture created these lines between 400 and 650 AD based on the carbon dating of the age of some sticks left mounted at the ends of some the lines.[71] This age is only a guess and perhaps should be based more on a much older time when climatic conditions turned this plateau into a very dry desert.

This paper will simply hypothesize a very plausible storyline for these Nazca Lines. Quickly changing climatic conditions due to one of Nemesis's crossings caused severe drought and harshened conditions for survival for the existing culture. These people obviously knew of many animals in their region that perished or were forced to migrate and no longer exist at this location. This ancient culture had stories handed down from verbalization that their Gods or nurturers from past times would return from the sky to aid them in these beleaguered times. Their idea was to communicate to the Gods or travelers in the sky and plea for their help. This culture knew of orcas and fish in the nearby ocean, flying birds, animals of the various terrains, small creatures such as lizards and spiders, but were engrossed by some myth or story that led them to worship special beings coming from skyward. The suspicion that flying machines was engrained in the minds of the Nazca culture of whatever age is difficult to dispel even for the debunkers of the Von Daniken claim.

XXVII.

WHO WERE THE BUILDERS OF THE PYRAMIDS AND THE WORLD'S OTHER MEGALITHS?

One of Von Daniken's heralded claims is that the Egyptian pyramids were impossible to build with the known tools and knowledge of the existing ancient culture unless there was intervention by some peoples with advanced technology such as ancient astronauts. An advanced technology possibly not even known today was required to design, plan, cut, transport, lift, and assemble with precise overall dimensions and alignments in a specific range of time of 20 years. Man's present construction equipment and methods would be severely tested if presented with the task of building just the Great Pyramid of Giza also known as the Pyramid of Cheops or Khufu. This fantastic structure was supposed to be built as a tomb for the Egyptian ruler, Khufu, within less than 20 years during his reign which does not include other projects of major structures such as two mortuary temples, a raised cause way connecting the temples, mastabas or tombs for nobles, large boat pits presumably that held boats, and three smaller pyramidal tombs for Khufu's wives.[71] The ancient Greek historian, Herodotus, writes that just the Khufu pyramid required 10 years of planning and 20 years of building. As an historian, Herodotus was sarcastically referred to as "the father of lies". His reliability of Egyptian history is criticized due to inadequate sources.[72] Initial studies revealed that the number of slaves required was 20,000 to 25,000. Later research now suggests that only a mere 6700 highly skilled artisans were required to complete the project.[73]

The entire focus of intense planning and laborious industry for this civilization with most of its populace was to build tombs for the nobility for their questionable after-lives. And, please hurry before they die. It sounds like science fiction from modern day writers. It all sounds like the present-day hadron collider project costing billions of dollars to find the illusive Higgs boson particle. So, when one discovers the Higgs boson, then what? When Khufu and his wives reach the after-life, then what? How will all the other mortal minions of the fourth Egyptian dynasty benefit? Either an amazing marketing campaign occurred perhaps based on the populace having jobs that assured their survival in this world or some over-riding fear drove these Egyptians. The fear was perhaps that the Pharaoh or these Gods on Earth had to be saved and preserved to protect the good life here on Earth. If that storyline can be sold, the elite certainly do benefit; they receive all the perks and riches while keeping their keepers fully occupied and away from political upheaval.

However, another line of thought comes forward. What if a well-advanced technological society had these pyramids and other monolithic structures built to serve as some kind of machine such as harnessing energy or as beacons for space travel? Then on one of Nemesis's visits this global society was destroyed leaving behind only their monolithic structures, some surviving ancient stone reliefs and texts and meager knowledge reduced to mostly verbalization which became our beloved myths. Then the likes of Pharaoh Khufu, his forefathers and other peoples around the world seized and re-used these monuments. During their exploitation, important archaeological evidence was falsified or destroyed or used to enhance the new pseudo rulers' control. After some musing and "food for thought" discussion let's proceed to examine some of the facts.

Let's review in detail the unbelievable accomplishments of the Egyptians in planning and building the Khufu or Cheops Pyramid, the highest structure in the world for 3800 years and still the most massive, having a volume of rock of 2,500,000 cubic meters. The pyramid is estimated to have 2.3 million blocks with an average opening in the joints of only 0.5 millimeters or 1/50th of an inch wide. As quoted from Wikipedia's, the Great Pyramid of Giza,[72] given the incredible accuracies and alignments for this structure, "… the four sides of the base have an average error of only 58 millimeters in length. The base is horizontal and flat to within ±15 mm (0.6 in). The sides of the square base are closely aligned to the four cardinal compass points (within four minutes of arc) based on true north, not magnetic north, and the finished base was squared to a mean corner error of only 12 seconds of arc. … the ratio of the perimeter to the height of 1760/280 cubits equates to 2π to an accuracy of better than 0.05 % (corresponding to the well-known approximation of π as 22/7)."[72] [from Egyptologist, Sir Flinders Petrie] Petrie related the precision of the casing stones as to being "equal to opticians work of the present day, but on a scale of acres."[72] How are all these technological and mathematical feats possible in 2560 BC when being so close to man's rise from the Stone Age ending about 3000 BC?

According to the written records, the 4th dynasty lasted from 2650 to 2480 BC when the better-quality pyramids of Giza were built. The pyramids were built during the reign of six pharaohs of the 4th dynasty: Huni (predecessor of Sneferi in the 3rd dynasty), Sneferi, Khufu, Djedefre, Khafre, Menkaure and the sixth and last, Shepseskal.[74] All other pyramids of the Nile Valley that preceded or followed this era were of poor quality, mostly in collapsed ruins, lessor flank inclines, and made from much smaller stones. Did the previous rulers and rulers that followed this 4th dynasty lose the art of pyramid building or never really possess this technology and tried to duplicate what they saw on the Giza plateau? Did the rulers of the 4th dynasty really evolve this technology in the span of about 170 years between its six known rulers or was there some divine intervention? Perhaps these Pyramids of Giza are much, much older going back beyond the time of the recorded Great Deluge of 11,500 years ago and claims of their origins are merely the seizure of these monuments that were then adorned with hieroglyphic reliefs and smaller temples and mastabas which produced the current dating developed by archeologists. Of course, this is speculation by this paper to try to explain and understand the following even more unbelievable construction data for the Khufu Pyramid.

Herodotus and Diodorus, two ancient Greek writers, interpreted and translated the stories of these reigning Egyptians 2000 years later as best they could by providing their spin to excite the readers of their times. Remembering the huge building projects of the 18th dynasty Pharaoh Akhenaten who reigned from 1351 to 1334 BC, these writers evaluated him along with Pharaohs of the famous 4th dynasty as being megalomaniacs. These rulers ruthlessly exercised their delusional fantasies of power and omnipotence over the entire populace. This type of dictatorial rule went against the contemporary philosophy of the ancient Greeks of that time. Herodotus particularly wrote about information he received about the building of the Great Pyramids of Giza; of course, this information could have been greatly distorted and may have been given for a much lesser pyramid that was actually attempted from the start. Herodotus quotes the Khufu Pyramid as taking 20 years; this time span is generally used by modern engineers and archaeologists for basing possible construction methods. Perhaps Herodotus did some napkin calculations to decide these rulers were absolutely obsessed and crazy.

Nevertheless, a time span was specified as were the knowledge, tools, and materials of that ancient period. The tools were crude surveying implements, hemp rope, and copper for bearing sleeves and saws, and boats, rails, rollers, sleds and levers made of imported woods from Lebanon. Bronze was not yet invented. Iron tools were supposed to be imported from the Hittites for working with the harder granites. The various stone products were quarried at numerous sites along the Nile River corridor as far as Aswan 500 miles away, the origin for the granite monoliths of the King's tomb. Of course, there were 20,000 to 30,000 slaves to perform the labor. A recent Egyptologist, Franz Lohner, devised a methodology using *rope rolls*, wooden brackets supporting wooden pins mounted in copper sleeves, to build this structure with an estimated more reasonable amount of 6700 skilled workers and artisans, not slaves.[74] His methodology eliminates the need for any type of ramps which were disputed as being impossibilities. Some ramps required as much stone material as the pyramid itself. Lohner claims to have tried his technique with equivalently heavy stones, about two to six tons being removed and dragged up an almost equivalent ramp of 40 degrees (less than the 52-degree incline of the pyramid flank) from an actual quarry using the aid of modern construction equipment; the required manpower was not available for his test. He never experimented with the 60-ton monoliths used for the King's tomb which could have been lifted on a lesser incline of the Grand Staircase. To date, his method is thought to be the most plausible; part of a rope roller tool and sleds were discovered and are displayed in museums.[74]

Franz Lohner and his associates are to be lauded for achieving the partial quest of solving the long-standing mystery of the construction of Khufu's Pyramid. A CAD/CAM video was created to show most of the details for quarrying, shipping by boat, transporting over land, finishing dimensions in the building yard, hoisting on the pyramidal flanks, positioning the blocks on the pyramid plateau, and the very special attention required by the harder granite 60-ton monoliths of the King's tomb. These special stones had to support the weight of rock of the remaining top third of the structure; this civil engineering design feat is totally remarkable and unbelievable. This paper does question rope rolls and hemp rope dragging 2 ½ ton stones up a 52-degree ramp as is proposed by Lohner. Hemp and sisal ropes even made in more modern times were excellent tools for sailing, tents, and canopies, but not for handling large loads. Ropes of this material continually pulling large loads over rope roll pins estimated to be 14 mm in diameter would quickly strain the rope causing frequent breakage. Industrial loads for lifting and controlling on inclines were kept small in the 1800th century due to rope failures until the invention of steel cables. More research is still required by Lohner.

Let's now continue with Lohner's calculations to achieve the construction in less than 20 years. These figures are staggering and even more unbelievable for any modern construction firm to achieve. To dedicate this many people to a single project for twenty or more years with no tangible benefits being accrued for the general populace is not even possible with modern governments and infrastructures. The pyramid is estimated to have 2.3 million blocks each weighing 2.6 to 2.9 tons on average. Blocks on the lower tiers are 6.5 to 10 tons each. If 500 stones were laid per day for 290 days per year for 20 years that total would equal 2,900,000 stones. Based on a 10-hour day, the rate of delivery is about one stone for every minute.[73] [74] This rate of flow of blocks needs to continue from cutting and removing the blocks from various quarries, transporting to the barges, loading on the barges, sailing to the Giza site, unloading from the barges, transporting to the site, finishing the stones, etc. These estimates do not include the granite workers and movers, the design and placement of ventilation shafts, tombs, grand staircase, underground chamber and entrance shafts. It seems that the ancient Egyptians, given their implements and materials of their day, were more advanced in the institutions of governing large populations and administering prodigious long-term projects than modern man is today.

XXVIII.

ACCOUNTING FOR POSSIBLE VERSIONS OF MAN'S HISTORY

Von Daniken's claim of ancient astronauts or an advanced, destroyed civilization cannot be necessarily ruled out. Various versions of man's history can be listed without embarrassment as being speculation based on some plausible inductive reasoning; no version can be fully proven. But, by simply listing these versions some sense of man's past can be assembled. Please use an open mind that is controlled by all the ideas and facts previously presented.

1. The Egyptian leadership and that of other ancient cultures was very enlightened and could achieve great engineering feats with the given resources. This leadership either came from an improved gene pool of earlier evolved Homo sapiens that were mostly destroyed by global cataclysmic events. The majority of surviving hominids either lost many of their special genetic traits or started over again with more primitive brains. The leadership was severely challenged by dealing with a less educated people and slowly perished due to inbreeding conflicts with a larger, more unruly population.

2. The ancient rulers were either a residual population of aliens from extraterrestrial sources or a more talented group of offspring produced by the merger of Earth's hominids. They may have had knowledge of anti-gravity technology that could move large objects. These rulers or mixed breed did not mix well with the original hominids and clashed with each other soon becoming extinct. The original aliens returned to their home planet orbiting Nemesis or fought each other and perished or tried to escape another catastrophe due to their visiting Nemesis star and failed.

3. The ancient rulers really did evolve from normal Homo sapiens and developed an unbelievable amount of technology and infrastructure in a short period of time. The rulers' megalomaniac characters created great engineering feats with

the tools, materials, and climatic conditions at hand. Characteristically, as with all of mankind's civilizations the Egyptian dynasties finally decayed and withered away. Perhaps their particular demise was due to plagues, drought, volcanic dust, floods, etc., brought by a previous visit of Nemesis around 2300 BC. It was during these times that plagues, drought and famine assaulted Egypt.

4.   The preferred version of this paper comes from ideas found and translated in the "Epic Tale of Creation" by the Sumerians and the Noah flood epic. The Sumerian epics of 6000 BC origins taught stories of creation handed-down by much earlier peoples several thousand years ago. The flood epic taught that much fauna and mankind's civilizations were destroyed by one of the worst cataclysmic events on one of Nemesis's visits 11,500 years ago. Mankind had developed a highly advance civilization globally that produced most of the monoliths and pyramidal structures prior to that time. These structures were machines with some unknown technical application. Surviving peoples after this global destruction lost most of their knowledge except for those memories passed on by verbalization and by the more rugged monuments and text that remained intact. The human survivors latched onto these monuments developing settlements next to them hoping to learn how to use these abandoned structures again. Many of the written language versions of history were embellished since any connection to true science was lost. This version does not require the arrival of any extraterrestrial beings.

5.   Another version comes from translations and interpretations by Zecharia Sitchin of several ancient Middle Eastern languages. Sitchin determined that aliens from a planet that comes through the inner solar system every 3600 years settled on Earth with a greater evolved knowledge and made a cross-breed, the modern Homo sapiens, from the existing hominids into their image to help with their labors on this planet. These aliens built most of the monoliths and pyramids for various technological reasons and had the means to transport themselves through the air. Sitchin does not make it clear what happens to the pure-bred aliens; one can assume that replacements stopped coming, and they eventually died or went home (do not rule out teleporting) leaving behind their cross-breeds to manage affairs after the cataclysmic destruction of Earth's surface by Nemesis's visit 11,500 years ago.

X.   This last version, labeled *X*, buzzes on the internet, movies and TV: the aliens known as the 'deep state' are still with us. They appear as any other normal person and are discrete about their actions; or, they are hidden in subterranean bunkers and move above the surface discreetly in UFOs. The *men in black* try to confuse or silence normal people who discover their presence or observe their UFOs. The *men in black* are given higher authority than any existing governmental official and act in the aliens' behalf. These *men in black* may also be liaisons between the aliens and the highest governmental and military leaders. These aliens or their representatives, "The Greys", are certainly prepared for another visit by the Nemesis Destroyer by living in deep, secretive, and sheltered subterranean caves.

XXIX.

THE CONCLUSIONS, OR SIMPLE TRUTH,
OF THE PAST 20,000 YEARS

Each version has great potential as a science fiction story or movie. However, what is learned is that no version can actually be proven – even version #3, now accepted by scientists and archaeologists, is not truly proven. What version is true or officially accepted by academics and governments makes little difference. What version you personally believe makes little difference. **What is paramount are two ideas taken from the following:**

1. from the studies of past geological and biological data

2. from cyclic data of climates and man's wavering and failing civilizations

3. from the worldwide, consistent storyline of myths and traditions

4. from the Earth's magnetic field changes and/or possible partial pole shift that can explain dramatic extinction events

5. from interpretations and translations of ancient texts and epics

6. from the primitive rock art depicting high energy plasma discharges between unknown celestial visitors seen in the ancient skies

7. from massive electrical arcing and sputtering as witnessed by mysterious landscapes on Earth and by space probe data recording Mars' surface and other orbiting bodies in the solar system

8. from the most reasonable source for asteroids and comets being the residual debris of electrical arcing of rocky surfaces and of collisions of asteroids

9. from the electrified environments that cause comet's tails, the helio-magnetosphere and the planetary magnetospheres that deflect the electrified solar winds

10. from the incredible, almost unbelievable and very questionable projects of monolithic structures created by ancient civilizations possessing only limited resources, tools, and knowledge.

THESE TWO IDEAS ARE THEN INDISPUTABLY PRESENTED:

Idea 1: A very technologically advanced globally-connected society existed to create the consistent traditions and monolithic structures that exist on every continent. Their technology has been lost or long forgotten for unknown reasons. The age of these structures is mostly given as the dating of skeletons and artifacts of organic nature or the dating of melted metals used as tools of a certain period. These dating methods cannot conclusively date the construction period of the monoliths since these archeological sites could have more than likely been contaminated by later ancient peoples who seized them for their own purposes, but had no responsibility for creating the original megaliths. The monolithic rocks themselves cannot be dated. If these worldwide societies of builders actually produced entire megalithic pyramidal structures from scratch, then why did all these societies disappear leaving very little or no trace as to how their monuments were built? Modern day scientists and engineers are still puzzled about how and why they were designed and erected. The dating of these structures with their 60 to greater than 1000-ton megaliths (some are granite requiring iron tools) could easily be either what is now proposed or of a much earlier period prior to the proposed cataclysm and Deluge of 11,500 years ago. Archaeologists today place all their bets and written theses on the currently accepted dating of events which includes the building of the complete structure during a few rulers' reigns. The dating is not necessarily certain and could go back several more thousands of years for these structures. The minds of scientists and historians must stay open-ended as should be the dating methods.

Idea 2: From the previously listed evidence it is also apparent that global cataclysmic events occurred to seriously affect mankind's evolution of his knowledge and his civilizations. Evidence shows that there was cyclic global destruction of varying degrees through the millennia going back 20,000 years and more. Erosion, land wasting, earthquakes, dust falling from the atmosphere by volcanism and the rotting of organic materials erased the most reliable records of Earth's surface and climatic history beyond the Neolithic Period. Any of man's existing civilization of prior times was completely erased except for the most rugged artifacts which are believed to be the existing monoliths. The better resolution of boundaries of unusual climatic, geological, archaeological eras stopped after the Mesolithic Period (8500 BC) of the Holocene Epoch. Some very good resolution into the Pleistocene Epoch led to identifying the boundaries of the last three stadial periods of glaciation taking recorded climatic changes back 20,000 years. The very well-defined boundary at the end of Pleistocene Epoch is the Younger Dryas Period which provides the best marker for one of cyclic changes caused by Nemesis. The identified cyclic changes occur about every 3600 years as is suggested by Zecharia Sitchin and this paper. *Whatever the exact cyclic period is not too important. What is important is that there is a highly probable cyclic period which can only be caused in classical physics by an orbiting Nemesis planet or star.* And, since electric arcing and sputtering is witnessed on the inner solar system bodies, these events strongly suggests the passing body is a star with strong magnetic and electrical properties and not a planet. Original planets or captured planets may still orbit this Nemesis star.

XXX.

PREPARATIONS FOR THE NEXT COMING OF THIS DESTROYER GOD, OR NEMESIS

Should man prepare for the next coming of Nemesis? Ancient peoples of both developed civilizations and primitive societies thought that Nemesis, a strange vision in the sky, was either a competing God to our Sun or Moon or was an evil Devil or Dragon. Could some urbanized groups of ancient astronauts or an advanced society of Homo sapiens possibly have known of their uncertain fate or revelation? This random fate depends on what happens each time as Nemesis passes through the inner solar system aligning itself in different ways with the other celestial bodies of both stars. If any previous advanced civilization was not prepared, then mankind's accumulated knowledge and infrastructure would be practically, totally destroyed. Then all the survivors would be reduced to the archaeological conditions of the early Neolithic or Mesolithic Ages. The survivors would have to surmount the ladder of knowledge once again and plod through the Bronze and Iron Ages trying to make sense of the remaining standing megaliths and pyramids. Their memories through verbalization would be passed to each succeeding generation with a consistent storyline of a horrible catastrophe. These storylines would become the worldwide myths that modern man has determined to have no technical basis.

If man learns about the knowledge of Nemesis and is convinced of another pending cycle then the answer is an unequivocal yes. Society should start as soon as possible to prepare and avoid another backward slip into the Stone Age or total extinction. But science or a convincing prophet has to come forward to warn about this revelation that Nemesis will bring. Teaching compassion and love for each other will also be required before it is shown how society can take such cooperative measures. And, what are these measures?

The very first appreciation is that the majority of Earth's population will likely perish either through direct annihilation or from the lack of sustenance once provided by the destroyed infrastructure. If the alignments are favorable and any close encounters or collisions are minimized then Earth may survive Nemesis's next coming with little affect. More than likely, one's fate will depend on the properties of one's location, such as the hemisphere or continental location; the proximity of oceans and large lakes; elevation above sea level; the closeness to volcanoes and fault lines, the path of monster storms and dust clouds, the survival of infrastructure in the region, etc. The anger of Nemesis is completely random and favors no particular human group or living species. Two major paths of preparation require consideration and hopefully both are eventually pursued. Perhaps the first path is already being pursued secretly.

XXXI.
PRESERVATION OF KNOWLEDGE BASE AND CURRENT TECHNICAL ADVANCEMENTS

The first type of preparation is to preserve our knowledge base and save certain groups of people represented by the complete diversity of the Homo sapiens gene pool or races who are very educated, compatible, healthy and youthful. These groups would be sheltered inside subterranean caves having the necessary self-supporting modern infrastructure. These underground cities located on each continent would have all the necessary elements for survival for perhaps five or more years. All the domesticated plants and animals would be housed in these caves, not too unlike the Noah epic. When the Earth recovers partially these preserved peoples of all races can re-emerge on the surface to start over again but with all modern knowledge and compassion for each other intact. There would be no misunderstanding of man's genesis story; and surviving peoples will know the value of cooperation and harmony. The true devil or evil would be known as the brown dwarf star, Nemesis, and not any evil residing within us or not the existence of an overbearing, punishing, fictitious god or gods. The evolution of life and man's knowledge would continue uninterrupted. The selection process for manning these subterranean cities would be similar to selecting leaders and crews for large aircraft carriers or nuclear submarines of today. Of course, there would be a periodic rotation of assignments to these mega-bunkers. The basic difference would be that every attempt will be made to mix the races within each city to prevent any further race and ethnic prejudice. The mixture of races also allows for less inbreeding, if part of each race is kept pure over time.

XXXII.

ASSURING THE BEST CHANCE OF SURVIVAL FOR THE MAXIMUM NUMBER OF PEOPLE

The second type of preparation is to assure as much as possible that the maximum amount of surface infrastructure survives to service survivors on the Earth's surface after any major assault by Nemesis. Politically and economically this is the most difficult shift in man's thinking that can be made. Perhaps earlier advanced societies knew the issues that certain infrastructure would create with each visit of Nemesis and completely avoided them. A listing is now composed for changing, reducing, or even eliminating the following "soft infrastructures". The order of presentation does not necessarily represent their degree of importance.

1. Some of the more important magnetic media needs replacing with *hardened* documents such as inscribed stone or at least high quality, long-life paper. Otherwise, electromagnetic storms will destroy all our digitized documents that exhibit our videos, pictures, music, financial records, and all other written materials.

2. Current forms of communication such as computers, the internet, phones and satellites will be useless. More hardened forms of communication will have to be resurrected and/or retained. These types of communication are normal mail systems, telegraph and shielded radio systems that can still partially operate during any intermittent but severe electromagnetic storms.

3. Some proportion of all commercial transportation systems will require backup hardened control systems that do not rely on computerization. These backup systems are especially true for rail systems and ocean transportation. A destroyed highway system and fuel distribution system will require the use of bicycles and horses and high-quality shoe ware.

4. Hardened power grids and pipe lines are required. The standard power distribution system with transformers will be systematically destroyed. Power

should be supplied by much smaller grid systems with lower voltages supplied by the source generators. More and more communities and individuals should install or have available their own generators that can be run with local natural gas or other readily available fuel. Major pipelines will require more automatic, passive shut-down systems due to random upheavals of the Earth's crust.

5. All major construction of infrastructure along shorelines and high-rise buildings should be reduced or stopped. Construction funding should be re-allocated toward in-situ underground, self-sustaining cities.

6. All construction of dams should cease and present dams should be drained and abandoned, especially the larger ones whose failure would wipe out in its path of destruction some of the largest cities on the planet.

7. All construction of nuclear power plants and nuclear materials processing plants should cease. All remaining plants should be decommissioned and their spent fuels stored until their half-lives become reasonable to inter them into the most rugged and reliable concrete casts. All present nuclear power plants have no passive design; they all require being connected to the power grid when shut-down. Without power, the spent fuel or any loaded reactor will go critical and release deadly radioactive materials into the atmosphere and water table for centuries. This verdict was established with the recent Japanese Fukushima disaster. The military-industrial complex still feels the need for nuclear arms as an equalizer similar to the six-shooter of the American Wild West. The six-shooter infrequently killed some innocent bystanders. But, this nuclear equalizer can kill the entire human race. The evolution of Earth's protective atmosphere and rocky crust have protected life against harmful radiation lurking both in space and the Earth's interior. Man should trust such matters to billions of years of evolution. Let's begin to sequester nuclear materials now and save Earth's fauna and flora, including ourselves.

8. The electromagnetic field energies created by the Earth being between the alignment of both the Sun and Nemesis is totally unknown. Perhaps some more knowledge about observed brown dwarfs and laboratory experiments may lead to a good estimation of what the highest field values will be. If the values are very high and are not deflected by the Earth's weakened magnetic field, then this radiation can be detrimental to the working of the human brain. To mitigate this issue of our brains being fried, the government could provide specially designed radiation-shielding helmets for each person on the surface. This

proposal sounds a little crazy, but not quite as crazy as the development of civil defense shelters during the Cold War with its accompanying nonsensical increase in production of nuclear tipped missiles having a human kill ratio of many more times the affected populations. The difference of this new proposal is that man is dealing with self-preservation and not self-destruction.

This mini-handbook for human survival also serves to protect us from other calamities predicted by modern astrophysics and planetary science. Some predicted plausible calamities from the field of astrophysics are: a sudden slight fluctuating expansion of the Sun's diameter producing momentarily higher electromagnetic radiation; a large Near-Earth-Object (NEO), comet or asteroid colliding with Earth; other comets or asteroids being perturbed by the gravitational attraction of nearby planets into changing orbits that make them NEOs; nearby gamma ray bursts (origins are still debated); and nearby supernovae that emit dangerous gamma ray radiation. Predicted calamities from the field of planetary sciences are: the loss of Earth's magnetic field and/or the ozone layer that protects us from ultraviolet and other dangerous radiation from the Sun; large continental earthquakes and movement of fault lines creating dam ruptures and total disruption of urban infrastructures; large underwater earthquakes or landslides causing tsunamis; mega-volcanoes causing huge dust clouds that blanket large areas or place climate-changing aerosols into the atmosphere; oscillating oceanic thermoclines that can cause dramatic climate change; and ice cap meltwater that raises sea level. Can one ever be safe from any of these calamities? Never. But, man can prepare now for some optimal survival modes in the hope of continuing his longevity along with some of his certain favored or domesticated flora and fauna. Also, man's overall planning of future infrastructures should perhaps start accounting for these calamities along with any coming of Nemesis.

XXXIII.
HOPE, FAITH, AND PRAYERS WILL PERSIST

The purpose of this book is to perfect the best marriage between myth and science. Predicting encounters with the Sun's sister star, Nemesis, does create this marriage, but at a cost. This prediction does, after all, require more evidence which, if proven, eventually may find its way into mainstream knowledge. The cost of this proving and admitting this prediction is the projection of a future of worldwide doom and resulting feelings of depression. But man is known to overcome all types of oppression and depression. The proposed preparations are probably politically impossible but still plausible. If worldwide governments do admit to the coming of Nemesis and decide to begin preparation and cooperate in the very near future of the next hundred years, there probably is enough time. Man's governments and other institutions could utterly fail at preparations if people become convinced of a pending doom and are not willing to sacrifice for the greater good of the species' survival. We all can hope and pray for one of the possible good outcomes: either good preparation or no danger happening from Nemesis's next coming or any other major calamity. Eternal hope is the test of everyone's faith and everyone's individual responsibility. Maybe the construction of the pyramids was a testament to man's faith and hope in past times. Here are some future hopes:

- This paper is only another fable or some science derived from bad assumptions that have some redeeming entertainment value and/or interesting ideas.

- Nemesis has been perturbed into a hyperbolic orbit never to return.

- Nemesis has been perturbed into a higher inclined orbit that has little chance of creating more disturbances of the planets or the Main Belt of asteroids.

- Nemesis, along with other celestial bodies, has greatly reduced its past stronger magnetic field thereby causing less damaging interactions.

- Nemesis has been perturbed into a changed aphelion that safely crosses between the outer planets' orbits with little effect on the inner solar system.

- The presence of Nemesis in our star system, for a various combination of reasons, presents no danger to Earth and its inhabitants for the next several million years. If certain sets of species of dinosaurs lasted for millions of years between mass extinctions per the fossil record, then why not man?

- Man does realize his history involves Nemesis and will prepare for its next coming. Man will continue his longevity and keep expanding his knowledge and wonder of the universe.

- The aliens that created Homo sapiens, if that is the case, will teleport all of us to a safe haven; or our souls are teleported automatically to another time and place to start over again if one's soul wishes. This mode of survival would really be a testament to trust and faith.

As you can readily sense, there is plenty of hope. So, keep the faith and continue to strive and learn. Our senses need both yin and yang for us to survive and this is good. All is good.

XXXIV.
APPENDIX 1: POSTULATED ORBITAL PATHS OF NEMESIS' BROWN DWARF STAR AND ITS PROPOSED PLANETS

Refer to Diagrams A and B for plan and side views of the orbital paths of both the Nemesis brown dwarf star's and the Sun's planets. The crossing orbital paths are postulated when Nemesis is at its periapsis or closest approach to the Sun. The diagrams are drawn roughly to scale using the unit of AU, if it is assumed that the expected orbits for Nemesis' planetary system do not exceed about 1 AU. In that case, no close intersection or crossing of orbits with the Sun's inner planets should occur. However, if the postulated *Dark Planet* or Nemesis's largest and most outer planet becomes perturbed, its new orbit may become elongated enough to cross Mars's orbit and come very close to Earth's orbit. Naturally, for that rare close encounter to occur, Mars or Earth would have to be fairly close to the Dark Planet's crossing. That coincidental meeting is highly improbable but is possible over thousands of orbital crossings. An orbital period of 3600 years as is postulated for Nemesis divided into the age of the solar system of 4.5 billion years is 1¼ million events. Other factors are also involved that can initiate the occurrence of close encounters, such as the incalculable interaction of numerous gravity fields and/or intense electromagnetic fields.

The diagrams do not show the possible close encounters that could occur with the Nemesis-star system as it comes close to the orbital paths of the outer planets: Jupiter, Saturn, Uranus, Neptune and their satellites. The large red spot on Jupiter, the ring systems around most to the outer planets, the inclined spin axes, captured satellites in various inclined and irregular elliptical orbits and the appearance of planetary and satellite surfaces being affected by high-energy electric arcing all indicate the possible effects of numerous close encounters with the electrified and highly magnetic Nemesis system that repeatedly returns to the Sun's inner and/or outer planetary system during its periapsis.

A diagram of NASA's proposed Nemesis perturbing the comets of a non-proven Oort Cloud of comets is presented in the diagram, Catastrophe I. The favored version of cyclic catastrophes for the author of this paper is Nemesis which provides a small enough cyclic period and ample opportunity to cause the numerous scarring of most of the terrestrial planets, many explored asteroids, the Earth's Moon and many of the outer planets' satellites. There is no question that science must eventually provide a hypothesis for why high energy electrical discharges have occurred so many times and in so many locations in the solar system.

A. Diagram A – Plan View of Orbital Paths with Nemesis

B. Diagram B – Side View of Planetary Orbits with Nemesis

C. Catastrophe I – Comet Storm Proposed by NASA

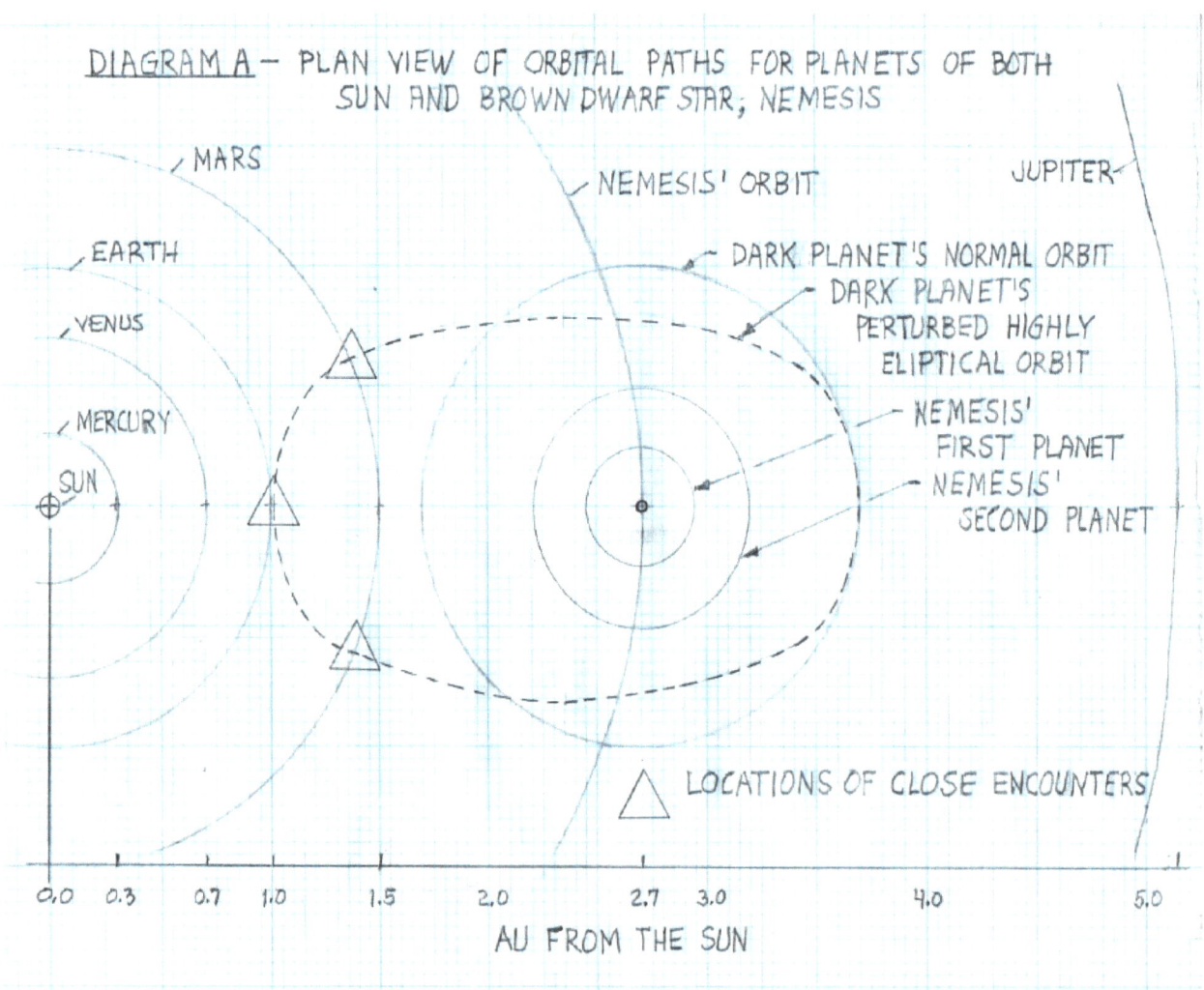

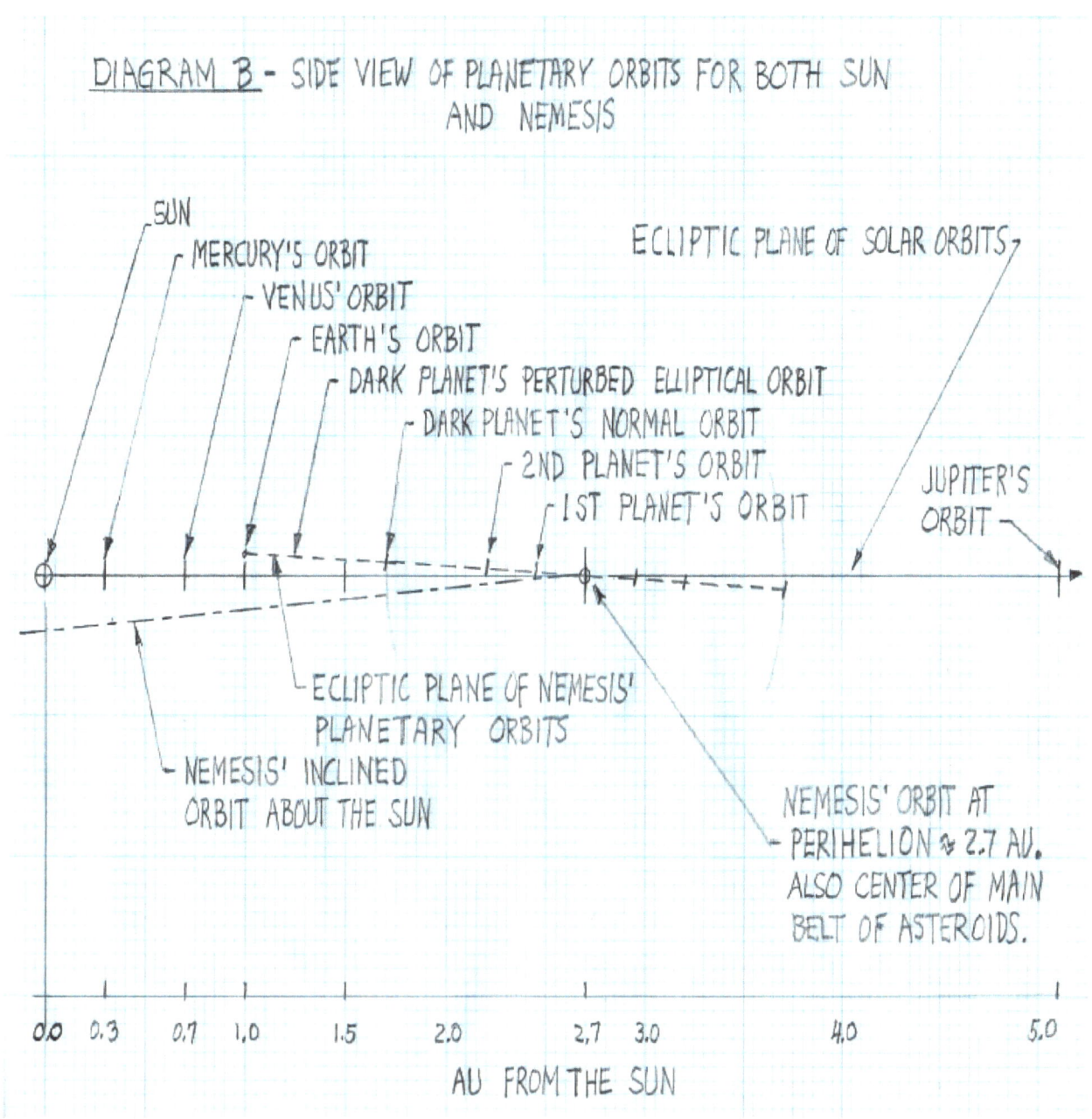

DIAGRAM B - SIDE VIEW OF PLANETARY ORBITS FOR BOTH SUN AND NEMESIS

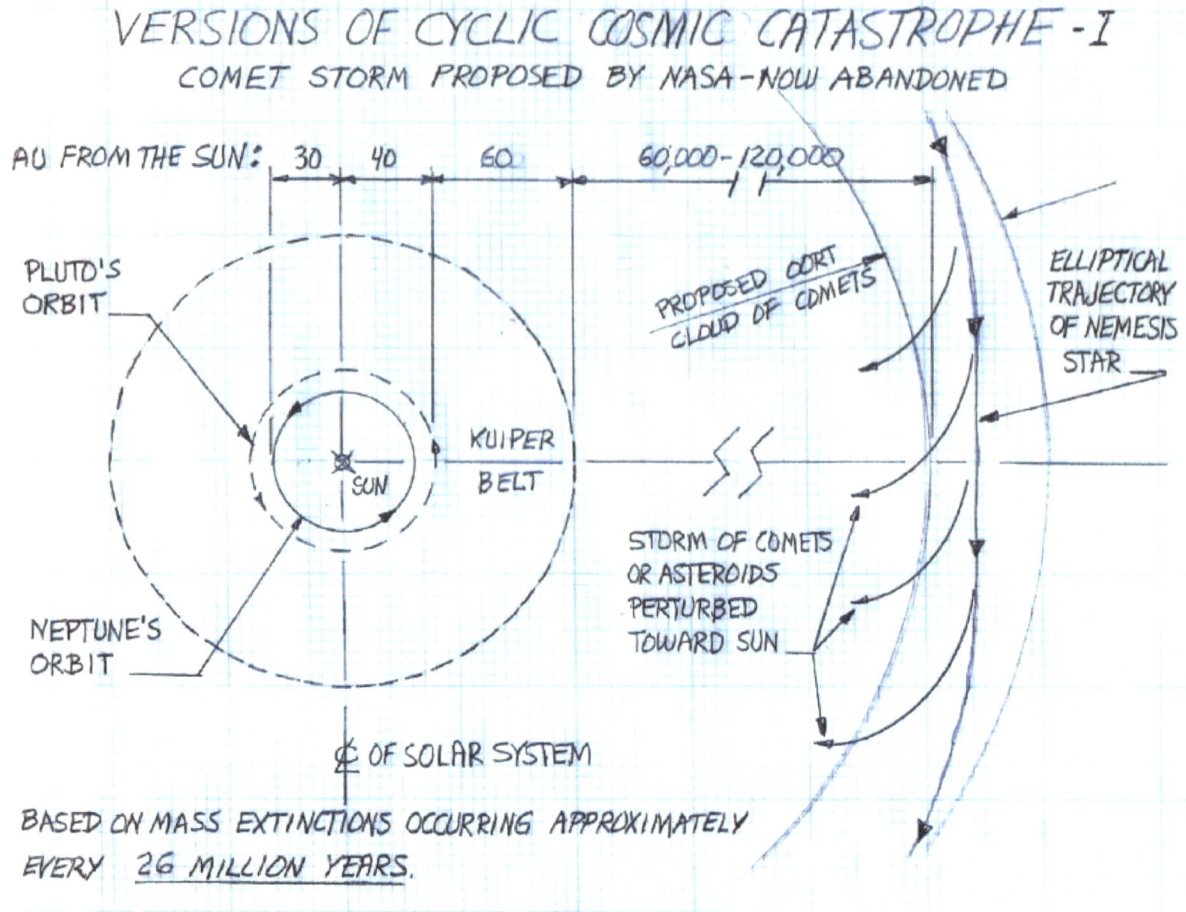

NASA had originally proposed periodic comet storms caused by an orbiting brown dwarf star's disturbing a cloud of comets orbiting the Sun at about 2 light years' distance. Unable to find a brown or red dwarf star at that proposed distance, and due to the times of mass extinctions in the fossil record not exactly fitting any possible orbital period of a Nemesis star, this hypothesis was officially abandoned.

NASA and the ESA need to consider a possible Nemesis brown dwarf star orbiting closer to the Sun and causing random chaos in the solar system and creating mankind's mysterious past.

XXXV.
ENDNOTES

SPECIAL NOTES

1. All Wikipedia references are dated as of 3/14/2014 unless otherwise stated. These references act as an old-fashioned encyclopedia when there is a strong need for obtaining data easily and quickly. Wikipedia does supply better cited works for those who require it.

2. This book's intent is to explore and connect many concepts in the various scientific disciplines to achieve a coherent storyline about catastrophism and hopefully create a paradigm shift. This book was never meant to be a peer reviewed journal.

3. Do not assume that any surviving authors referenced herein endorse each other or any of the new material presented in this book.

[1] Wikipedia; Nemesis (hypothetical star)

[2] Wikipedia; Nemesis (hypothetical star); claimed periodicity of mass extinctions

[3] Wikipedia; Nemesis (hypothetical star); past, current, and pending searches for Nemesis

[4] Wikipedia; Wide-field Infrared Survey Explorer (WISE); targets for the solar system

[5] Wikipedia; Vesta asteroid

[6] Wikipedia; Brown dwarf star

[7] Wikipedia; Wide-field Infrared Survey Explorer (WISE); Results

[8] Wikipedia; Teide 1 (brown dwarf)

[9] Wikipedia; Gliese 229B (brown dwarf)

[10] Wikipedia; Barycenter

[11] Wikipedia; Little Ice Age

[12] Wikipedia; Carbon14 with activity labels

[13] Wikipedia; Little Ice Age; dating

[14] Astronomical Timeline by Erling Poulsen; www.rundetaarn.dk/engelsk/observatorium/timeline.htm

[15] Wikipedia; 17th century BC (1/10/2014)

[16] A New Ancient History by Robert Schoch; www.petragrail.tripod.com/newhistory.html

[17] Wikipedia; 6th millennium BC; events

[18] Wikipedia; Sar

[19] Wikipedia; 4th millenium BC; environmental changes

[20] Wikipedia; Mesolithic

[21] Wikipedia; Neolithic

[22] Wikipedia; Geological Time Scale.png

[23] Wikipedia; Oldest Dryas; Dating

[24] Wikipedia; Post-Glacial Sea Level.png

[25] Wikipedia; Older Dryas

[26] Wikipedia; Bolling oscillation

[27] Wikipedia; Allerod oscillation

[28] Wikipedia; Five Myr Climate Change.png

[29] Wikipedia; Phanerozoic Climate Change.png

[30] Wikipedia; Quaternary glaciation

[31] Wikipedia; Vostok-ice-core-petit.png: NOAA

[32] Climate Variability on Millennial Time Scales; www.envsci.rutgers.edu/≈broccoli/..../millenial.ppt

[33] Wikipedia; Younger Dryas

[34] Wikipedia; Younger Dryas; Abrupt climate change, Global effects

[35] Wikipedia; Younger Dryas, Causes

[36] Wikipedia; Younger Dryas, Effect of agriculture

[37] Wikipedia; Northern Ice Sheet, hg.png

[38] D.S. Allan and J.B. Delair; Cataclysm! *Compelling Evidence of a Cosmic Catastrophe in 9500 BC*, 1997

[39] Wikipedia; Typhon (Greek god)

[40] Wikipedia; Ashur (Assyrian god)

[41] Wikipedia; Marduk (Babylonian god)

[42] Wikipedia; Erra (Akkadian god)

[43] Wikipedia; Nibiru (Sumerian God)

[44] D.S. Allan and J.B. Delair; *Cataclysm! Compelling Evidence of a Cosmic Catastrophe in 9500 BC*; Fig. 4.13, page 228

[45] White, John; *Pole Shift*, 1980

[46] Hapgood, Charles; *Earth's Shifting Crust*, 1958; and *The Path of the Pole*, 1968

[47] Farrand, W.P.; "Frozen Mammoth and Modern Geology" by *Science Journal*, 1961

[48] Velikovsky, Immanuel; *Worlds in Collision*, 1950

[49] Sitchin, Zecharia; *The Twelfth Planet*, 1976

[50] Ettinger, Douglas; www.ettingerjournal.com; The Earth's Metamorphosis

[51] Gideons International; Holy Bible, Genesis, chapter one

[52] Talbott, David and Thornhill, Wallace; *Thunderbolt of the Gods*, 2005; and *The Electric Universe*, 2007

[53] Talbott, David; *The Saturn Myth*, 1980

[54] Talbott, David and Thornhill, Wallace; "Symbols of an Alien Sky", 2011, produced for DVD and video streaming

[55] Wikipedia; Jupiter, the planet

[56] Wikipedia; Stardust (spacecraft); sample processing; results

[57] NASA JPL; California Institute of Technology; Stardust – NASA's Comet Sample Return Mission; Stardust Findings Suggest Comets More Complex Than Thought; Dec 14, 2006

[58] Talbott, David and Thornhill, Wallace; "Symbols of an Alien Sky"; Episode 3; *The Electric Comet*, 2011

[59] Wikipedia; Pioneer Missions

[60] Wikipedia; Mars; Geography and naming of surface features

[61] NASA JPL; California Institute of Technology; Mars Global Surveyor Mission

[62] Wikipedia; Mars Global Surveyor; Objectives; Mapping

[63] Talbott, David and Thornhill, Wallace; "Symbols of an Alien Sky"; Episode 2; The Lightening Scarred Planet Mars; 2011

[64] NASA JPL; California Institute of Technology; Mars Exploration Rover Mission, 2003

[65] Wikipedia; Scientific Information from Mars Exploration Rover Missions

[66] Von Daniken, Erich; *Chariots of the Gods*; 1968

[67] Wikipedia; Erich von Daniken; Critique and criticism

[68] Wikipedia; List of largest monoliths in the world; quarried, moved, and lifted monoliths

[69] Wikipedia; Vimana; Etymology and usage; In Sanskrit literature: Vedas and Ramayana

[70] Wikipedia; Nazca Lines; History; Purpose

[71] Wikipedia; Great Pyramid of Giza

[72] Wikipedia; Herodotus

[73] Lohner, Franz; *Building the Great Pyramid*, 2006

[74] Wikipedia; 4th Dynasty of Egypt

ABOUT THE AUTHOR

The Moon-Earth enigma has been a lifetime pursuit since the mid-80's for Doug Ettinger, born in 1944 in Pen Argyl, Pennsylvania. Doug graduated from Lehigh University with a B.S. in mechanical engineering and has retired after a 22-year career in nuclear engineering. He has two sons and currently lives in Pittsburgh, Pennsylvania, with his significant other, Rhonda Smith. Some of his main interests and distractions are snow skiing, hiking, kayaking, sailing, bicycling, tennis, ballroom dancing, movies, and chess.

Doug wrote his initial journals about the Moon enigma; NASA is still plagued with how the Moon and Earth became a system. His writings are divided into three hypotheses: Earth's Metamorphosis (EMM) hypothesis (about Earth's collision with a rogue planet and transferring orbits); Collocation of Stars and Planets (CSP) hypothesis (about how planets seek their orbits and stars find binary companions); and Supernovae Seeding (SNS) hypothesis (about how new stars and planets are birthed through the expulsion of supernova ejecta). His passion and self-training cross many fields of science including astronomy, astrophysics, planetary science, oceanography, physics, geophysics, and particle physics. Hopefully, his depth of knowledge in this wide spectrum of disciplines is adequate to at least meet the minimum of academic standards.

His initial interest started with reading about solar system formation and all the unexplained anomalies that exist. One of these anomalies, the Moon enigma, caught his attention after reading Isaac Asimov's claim that the Moon is actually a planet. Combining the anomaly of the asteroid belt led Doug to consider the ideas for his Earth's Metamorphosis hypothesis.

Doug helped his one son with his science fair project by providing a primitive computer program that could create your own stellar systems or modify the starting conditions of the solar system and then run the program for a given amount of time to determine the end conditions. His son's project was taken to a higher level of competition at a university. The judge explained that his criteria for collisions were invalid because the two-body problem and calculus can prove otherwise. His son did not know how to defend his position by explaining the limitations of the two-body problem. Several months later a comet broke up and crashed into Jupiter for everybody to see. These unfolding scenarios were the seed that finally launched Doug on his way to resolve the Moon enigma.

144

Since posting my original journals and some subsequent editions in 2013, many new developments and ideas have occurred. Also, some power point presentations have been included to aid teachers who wish their students to sometimes think outside the present box of paradigms.

Special attention was given to the Sun's sister star, currently named Nemesis, which at the time, was being hunted by space telescopes. I wanted better reasons for mass extinction events. The accepted notion of a very long period star orbiting the Sun and infrequently disturbing an imaginary Oort Cloud of comets was not believable or proven. NASA has recently revised or given-up their model. The disturbances of this star are now considered to be perturbations of the newly discovered Kuiper Belt objects. Doug slowly and cautiously became convinced that a much closer, very dim brown dwarf star orbited the Sun and intersected the orbits of the Sun's planets, causing occasional chaos. The underpinning of this idea originated with Zecharia Sitchin's book, *The Twelfth Planet,* that involved a planet with moons that orbited the Sun every 3600 years.

Doug adapted his idea by making Sitchin's intruding planet become a brown dwarf star with its own set of planets. This idea easily dovetailed into his postulation of "Earth's Metamorphosis" found in his original journal. The Earth's "Great Deluge" is linked with one of Nemesis's crossings of the inner solar system, which is well dated to be 11,500 years BP, at the end of the Younger Dryas geological period. This led to investigating other crossings at approximate intervals of every 3600 years in the journal "A Brief History of Mankind's Chaotic Past".

Subsequently, Doug discovered the ideas formulated by a group of independent scientists called the "Electric Universe". Their concepts unbelievably corroborated his journal by indicating an electrified and magnetic brown dwarf star periodically enters our solar system. This star, and/or its planets, interacts with the Sun's planets, producing scarring and giant pitting through gigantic arc discharges that are readily observable with NASA's space probes and telescopes. The "Electric Universe" also provided a convincing dialogue, via comparative mythology, that humans witnessed and recorded some of these events.

After taking an Alaskan cruise and talking with the indigenous people, called the Tlingit, I learned that their very distinct tribal cedar dug-out canoe docking systems are found not only at sea level, but at locations 100 to 150 feet above the ocean on the mountain side. The Tlingit flood tradition and this recognition triggered my next journal "The Great Deluge: Fact or Fiction". Doug combined the ideas of other books written about the Flood, including the exclusion of some of their incongruent arguments. He added the idea that one of Nemesis's planets had a close encounter with Earth that caused many calamities, such as a huge arc discharge that destroyed the Laurentide Ice Sheet and initiated the Noah Flood, which is found in other flood traditions throughout the world.

One of the difficulties for the "Electric Universe" is convincing consensus science that "charge separation" or flow of segregated electrons and electrical current flow through space do exist. This gave me ideas about how these phenomena are possible by establishing the idea of "electron asymmetry" in a journal by the same name. "Electron asymmetry" not only helps to explain "charge separation" and high energy arc discharges between close encounters of celestial bodies, but provides the reason for why gravity exists.

CONNECT WITH DOUG ETTINGER

Contact me at email: dougettinger@verizon.net
Learn and read other articles: EttingerJournals.com
Friend me at Facebook: facebook.com/Star-Planetary Origins

RELATED ARTICLES BY DOUG ETTINGER

- The Enigma of the Giza Pyramids of Egypt – When and How Were these Megaliths Designed and Built (ettingerjournals.com/dbe_giza.shtml)

- Corroborating Massive Solar Eruption Causing Catastrophism on Earth – Using Robert Johnson's Model for Providing Enough Energy to Cause Tectonic Uplift Processes through Thermal Expansion and Phase Change of Rocks – (ettingerjournals.com/dbe_solar_eruptions.shtml)

- A Brief History of Mankind's Chaotic Past – Post-Paleolithic Times (20,000 Years Ago) to the Present that embraces Nemesis' Influence and Prophecy – (ettingerjournals.com/dbe_mankind.shtml)

- Ancient Sun-God Descriptions Give Proof for an Orbiting Nemesis Star – A Comparative Study of Mesopotamian and Egyptian Sun-God Depictions Referenced in a Ev Cochrane Article (ettingerjournals.com/dbe_sun_gods.shtml)

- Problems with the Saturn Myth's Polar Configuration – Replacing the Proto-Planet Saturn Idea with an Orbiting Brown Dwarf Star (ettingerjournals.com/dbe_saturn_myth.shtml)

- Electron Asymmetry – What is this Asymmetry? How Does It Affect Our Universe? (ettingerjournals.com/dbe_electron_asymmetry.shtml)

- Earth's Metamorphosis (EMM) Hypothesis – The Event and Aftermath of Earth's Collision with a Large Impactor that Changes Its Orbit, Spin Axis, and Surface Features (ettingerjournals.com/dbe_emm.shtml)

READ THE BOOKS BY DOUG ETTINGER

The Great Deluge: Fact or Fiction?
smashwords.com/books/view/919340
(ebook)

The Great Deluge: Fact or Fiction Companion Book
smashwords.com/books/view/919342
(ebook)

The Great Deluge: Fact or Fiction?
amazon.com/dp/1797740695
(paperback)

www.ingramcontent.com/pod-product-compliance
Lightning Source LLC
Chambersburg PA
CBHW051017180526
45172CB00002B/385